John Powell
2490 Fauhall Rd
Kelowna
860 - 9696

Soil Geomorphology

Soil Geomorphology

An integration of pedology and geomorphology

John Gerrard

Senior Lecturer in Geography
School of Geography
University of Birmingham
UK

CHAPMAN & HALL
London · Glasgow · New York · Tokyo · Melbourne · Madras

Published by Chapman & Hall, 2–6 Boundary Row, London SE1 8HN

Chapman & Hall, 2–6 Boundary Row, London SE1 8HN, UK

Blackie Academic & Professional, Wester Cleddens Road, Bishop-briggs, Glasgow G64 2NZ, UK

Chapman & Hall, 29 West 35th Street, New York NY10001, USA

Chapman & Hall Japan, Thomson Publishing Japan, Hirakawacho Nemoto Building, 6F, 1–7–11 Hirakawa-cho, Chiyoda-ku, Tokyo 102, Japan

Chapman & Hall Australia, Thomas Nelson Australia, 102 Dodds Street, South Melbourne, Victoria 3205, Australia

Chapman & Hall India, R. Seshadri, 32 Second Main Road, CIT East, Madras 600 035, India

First edition 1992

© 1992 John Gerrard

Typeset in 10/12 Palatino by Pure Tech Corporation, Pondicherry, India
Printed in Great Britain by St Edmundsbury Press, Bury St Edmunds
ISBN 0 412 44170 5 (HB)   0 412 44180 2 (PB)

A catalogue record for this book is available from the British Library

Library of Congress Cataloging-in-Publication data available

Contents

Acknowledgements

A great many people have been instrumental in bringing this book to fruition; too many to name individually. However, there are a few who deserve special thanks. These are Caroline Olson, who provided very helpful advice and encouragement in the early stages, and Peter Knuepfer and Les McFadden, who, by inviting me to the 21st Binghamton Symposium on Geomorphology, created the opportunity for me to meet and discuss with all the important researchers in the field of soil geomorphology. Their help has been invaluable. Thanks are also due to the following authors, publishers and learned bodies for permission to reproduce figures and tables:

R.R. Arnett (Table 4.7); Figure 9.1 reproduced from *Glaciers and landscape* (D. E. Sugden and B. J. John) by kind permission of the authors and Edward Arnold Ltd; E. Bettenay (Figure 8.6); S. F. Burns (Figure 2.6; Table 2.1); B. E. Butler (Figure 12.5); Figure 3.4 reproduced from *Tropical soils and soil survey* (A. Young), Figure 12.5 reproduced from *Landform studies from Australia and New Guinea* (eds J. Jennings and J. A. Mabbutt), Table 3.1 reproduced from *The soil resources of tropical Africa* (R. P. Moss) and Table 11.1 all by kind permission of Cambridge University Press; C. J. Chartres (Tables 7.2, 7.3); R. J. Chorley (Figure 2.2); Commonwealth Scientific and Industrial Research Organization (Figure 8.6); A. J. Conacher (Table 4.7); R. U. Cooke (Figure 10.2); Elsevier Scientific Publishing Co. (Figures 5.4, 5.5, 6.1; Tables 4.5, 4.6, 5.3, 7.1); Elsevier Scientific Publishing Co. (*Engineering Geology*) (Figure 10.3); Elsevier Scientific Publishing Co. (*Geoderma*) (Figure 2.7, Tables 7.2, 7.3); A. E. Foscolos (Table 12.2); Geographical Society of New South Wales (Table 4.7); George Allen and Unwin (Figures 2.6, 9.4, 12.8; Table 2.1); R. D. Green (Figures 8.2, 8.3, 8.4; Table 8.1); J. R. Hails (Figure 6.1); Houghton Mifflin Co. (Figure 6.5; Table 5.2); R. J. Huggett (Figure 2.7); D. L. Hughes (Table 12.2); Figure 3.8 reproduced from *The geography of soils* (B. Bunting) by kind permission of Hutchinson General Books Ltd; J. D. Ives and the International Mountain Society (Figure 2.5); Figures 4.3, 4.4 and Table 4.2 reproduced from *Slopes: form and process* by kind permission of the Institute of British Geographers; International Society of Soil Science

(Figure 2.3; Table 4.3); Iowa State University Press (Figure 9.2); B. A. Kennedy (Figure 2.2); R. T. Legget (Figure 9.5); D. M. Leslie (Figure 12.9); Longman Group Ltd (Figure 3.9); W. M. McArthur (Figure 8.6); Figures 3.1 and 8.5 reprinted by permission from *Nature*, **138**, 549 and **196**, 836, © 1936 and 1962 Macmillan Journals Ltd; W. C. Mahaney (Figures 12.2, 12.3; Table 12.2); B. Matthews (Figure 9.3); R. B. Morrison (Figures 12.2, 12.3); R. P. Moss (Table 3.1); New Mexico Bureau of Mines and Mining (Figures 10.6, 10.7, 10.8, 10.9; Tables 10.5, 10.6, 10.7); the Editor, *New Zealand Journal of Geology and Geophysics* (Figure 12.9); M. A. Oliver (Figure 4.1); Oxford University Press (Figures 3.2, 3.3, 3.5, 3.6, 5.2, 6.4, 7.1, 12.6; Tables 6.2, 6.3, 6.4, 6.5, 12.3); Oxford University Press (New York) (Figure 12.4); Figure 9.5 reproduced from *Glacial till* (ed. R. T. Legget) by kind permission of the Honorary Editor, Royal Society of Canada; R. V. Ruhe (Figure 6.5; Tables 5.1, 5.2); N. W. Rutter (Table 12.2); Soil Survey of England and Wales (Figures 8.2, 8.3, 8.4, 9.3; Table 8.1); Springer-Verlag (Figure 8.1); P. J. Tonkin (Figure 2.6; Table 2.1); University of Chicago Press (Figure 9.6); A. F. Warren (Figure 10.2); R. Webster (Figure 4.1); Table 5.1 reproduced from *Soil Science*, **82** (6), 453, © 1956 Williams & Wilkins Co., Baltimore; the Editor *Zeitschrift für Geomorphologie* (Figures 2.4, 4.2, 5.1).

Preface

It is becoming increasingly clear that, in order to alleviate and control many of the world's major environmental problems, a thorough understanding of the workings of the Earth's surface is essential. It is also clear that aspects of the Earth's surface cannot be examined in isolation. One of the most fruitful interactions to study is that between pedology and geomorphology. When *Soils and Landforms* was published in 1981 there were signs that this integration – between pedology and geomorphology – was becoming respectable and that a new focus of study had emerged. Research and interest over the last ten years have shown this to be true. Numerous influential publications on the theme have appeared, such as Peter Birkeland's *Soils and Geomorphology* (Oxford University Press, 1984), *Geomorphology and Soil* (Allen and Unwin, 1985), *Soils and Quaternary Landscape Evolution* (J. Wiley and Sons, 1985) and *Soils and Quaternary Geology* (Clarendon Press, 1986). The culmination of this decade of research was the 21st Binghamton Symposium in Geomorphology, which took as its theme *Soils and Landscape Evolution*. Soil geomorphology has become fashionable. The aim of this book is to establish the foundations of soil geomorphology and to provide an update and synthesis of the more recent work. New chapters have been added, dealing with soils on river terraces and alluvial fans, and soils associated with desert and periglacial landforms. At the same time, all the other chapters have been thoroughly revised with the addition of a considerable amount of new material. Over 200 references refer to work published since 1980. I hope that the valuable collaboration between pedologists and geomorphologists, exemplified by much research in the last decade, continues in the same vein.

John Gerrard
1992

1

Soil geomorphology – the approach

Geomorphology and pedology are two of the more important disciplines in the Earth Sciences. The former deals with the arrangement and differentiation of landforms and the processes that have been or are shaping them. Pedology is concerned with the processes involved in soil formation. Until the last twenty or thirty years pedology and geomorphology tended to be treated separately with only a token awareness of the influence of one on the other and vice versa. There were, of course, exceptions and it is possible to identify overlapping theoretical aspects between geology and pedology in the evolution of the soil profile concept (Tandarich *et al.*, 1987).

Geomorphology was concerned essentially with producing time-dependent models of landscape evolution such as W.M. Davis's *Geographical Cycle* of landscape evolution. The form of the land was the major focus with little mention of process and scant attention paid to the soil and regolith materials. Admittedly Davis did emphasize the way in which a graded waste sheet might develop on slopes but there was little analysis of the way in which that waste sheet interacted with slope processes. However, Gilbert, in the late nineteenth century, was emphasizing the equilibrium between landforms, soils and slope processes. In Europe, W. Penck, while concentrating on the general elements of landform evolution, did at least, attempt to relate the progression of soil formation towards maturity with slope processes.

The change in emphasis in geomorphology in the 1960s and 1970s led to a greater concentration on surface processes and the short-term changes occurring in the landscape. Investigations of drainage basins and storm hydrographs demonstrated the influence exerted on these phenomena by the surface covering of soil and vegetation. It was recognized that many soil processes were also geomorphological processes and the distinction between geomorphology and pedology was being blurred at the edges of the disciplines. The development of the dynamic equilibrium approach in geomorphology also led to an awareness of relations between geomorphology and pedology. Dynamic equilibrium stresses that an equilibrium form may last as long as the controlling forces are unchanged (Hack, 1960, 1980). Within a single erosional system, a balance exists between the processes of erosion and the resistance of rocks and all elements of the topography are mutually adjusted so that they are downwasting at the same rate. In this framework soil assumes a major role. Often ideas of stability and instability can only be established with reference to the soil cover.

Modern research is increasingly demonstrating the close dependence of soils and geomorphology and a new discipline, 'soil geomorphology' or pedogeomorphology as proposed by Conacher and Dalrymple (1977) has emerged, incorporating traditional approaches to soils as well as modern soil engineering. These developments were forecast by Robinson (1949) in his introduction to *Soils, their origin, constitution and classification*, with the words 'the domain of pedology may come to engross a considerable amount of dynamic geology'. The impetus has been maintained by workers such as Birkeland (1974, 1984) and his many former research students, also by publications such as *Geomorphology and Soils* (Richards, Arnett and Ellis, 1985), *Soils and Quaternary Landscape Evolution* (Boardman, 1985a) and *Soils and Quaternary Geology* (Catt, 1986). The establishment of soil geomorphology as a major focus of earth science research was reflected in the theme of the 21st Binghamton Symposium in Geomorphology. The proceedings of this symposium, *Soils and Landscape Evolution* (Knuepfer and McFadden, 1990), is a statement of current work in soil geomorphology.

Soil geomorphology is basically an assessment of the genetic relationships of soils and landforms. Thus the emerging marriage of pedology and studies of surficial processes provides a critical framework for both the study of soil genesis and the study of evolution of and/or stability of landscape elements (McFadden and Knuepfer, 1990). There are many ways in which the integration between geomorphology and pedology is expressed. Geomorphological and pedological processes interact on hillslopes especially where the movement of soil and water is considered. Geomorphological processes may create distinctive landforms, such as erosion surfaces, which have a great influence on soil types and distribution. Usually, however, it is the creation of landform type and superficial materials in harness which is of greatest significance. Investigation of how landforms develop and of rates and results of processes has relied on the study of surfaces and sediments. But soils that have developed on geomorphic surfaces during and after surface stabilization have the potential to record more accurately the history of landscape development.

GEOMORPHOLOGY AND MODELS OF SOIL FORMATION

It is probably true that pedologists have always been more aware of the importance of geomorphology, especially landforms or topography. Topography has long been recognized as one of the factors influencing soil development and has been included in many models of soil formation. It is generally agreed that Dokuchaev in 1898 was the first to suggest a soil-forming factor equation. His equation was:

$$s = f(cl, o, p)t^0$$

where s = soil, cl = climate, o = organisms, p = parent material and t^0 represented relative age (youthfulness, maturity, senility). The equation should be taken as a symbolic expression or a conceptual model and does not imply that it can be 'solved' in a mathematical sense. Although relief or topography is not one of the factors, Dokuchaev did acknowledge relief as important, but chiefly in the formation of 'abnormal' soils. Later workers, such as Hilgard (Jenny, 1961b) and Shaw (1930), modified this synthesis and added relief to the soil-forming factors. This culminated in Jenny's (1941) state factor equation:

$$s = f(cl, o, r, p, t, \ldots)$$

where s denotes any soil property, cl is the environmental climate, o is animal organisms, r is relief, p is parent material and t is time since the start of soil formation. The important point to stress is that the factors are variables that define the state of the soil system. Jenny, in 1961, modified the equation to make it more applicable to modern thought concerning ecosystems. The revised equation is:

$$l, s, v, a = f(L_0, p_x, t)$$

where l is any property of the ecosystem in its totality, soil properties are denoted by s, vegetation by v and animal properties by a. L_0 represents the assemblage of properties at time zero, p_x, the flux of materials and t is the age of the system. The configuration of the system, such as its slope, exposure and topography, is a subgroup of L_0 as are considerations of the mineral and organic matrix of the soil. Climate is a subgroup of the flux potential p_x.

Five broad groups of factors in accordance with the five state factors, are suggested. These are:

$$l, s, v, a = \begin{cases} f(cl, o, r, p, t, \ldots) & \text{climofunction} \\ f(o, cl, r, p, t, \ldots) & \text{biofunction} \\ f(r, cl, o, p, t, \ldots) & \text{topofunction} \\ f(p, cl, o, r, t, \ldots) & \text{lithofunction} \\ f(t, cl, o, r, p, \ldots) & \text{chronofunction} \end{cases}$$

The dominant factor is placed first. A more convenient way of writing this, as suggested by Jenny (1961a), is:

$$l, s, v = f(r)_{cl, o, p, t} \quad \text{topofunction}$$

Soil formation was considered by Wilde (1946) as a dynamic process with interdependent soil environmental factors interacting through time, thus

$$s = f(G, E, B) \, dt$$

where G is parent material, E represents environmental factors and B represents biological factors. According to Simonson (1959)

$$s = f(A, R, T_1, T_2)$$

where A represents additions, R represents removals, T_1 represents transformations and T_2 translocations to, within and from the profile. Runge (1973) devised an energy model (Chapter 2) where water and organic matter were the organizing and retarding vectors of pedogenesis. Thus

$$s = f(W, O, T)$$

where W is soil energy related to leaching potential of water, O is organic matter production and nutrient cycling and T is time. Johnson and Rockwell (1982) produced a model similar to that of Wilde where

$$s = f(P, D) \, dt$$

where s is the strength of soil formation of an entire solum or a single attribute, P and D are two interactive sets of passive and dynamic factors of soil formation, dt represents the change in both passive and active factors.

The great problem with this type of approach is that it is extremely difficult to handle in a modern quantitative fashion. Although expressed in a mathematical way, these are really verbal models in the sense used by Dijkerman (1974). Yaalon (1975) has reviewed the attempts that have been made to solve these state factor equations. Lithofunctions present problems because of the difficulties in assigning numerical values to parent materials, however, the use of binary attributes might be profitable. Topofunctions are easier to manage and there have been many graphical and numerical attempts to relate soil properties to landscape elements, such as slope angle and position. Some of these are discussed in Chapter 4. One of the problems in fitting linear or curvilinear relationships to this type of data is that both slope angle and soil properties are strongly autocorrelated. This means that individual soil properties and gradient angles on single slopes are correlated with themselves and are not independent entities, which may invalidate many statistical findings.

There have been numerous qualitative studies of the effect of climate on soils and a number of general conclusions have emerged which may well be capable of quantification in the future. Most graphical solutions to date have been concerned with chronofunctions, but there is also a rapidly growing number of studies which have attempted to quantify the rate of change of particular soil properties. The major fact to emerge from these studies is that not only does the rate of change vary from one soil property to another but so does the form of the mathematical function. This is important when considering the vexing questions such as soil maturity and whether dynamic equilibrium, as a concept, can ever be applied to soils. Numerical solutions of true biotic functions are rare, but knowledge of the general role of vegetation is substantial. Thus progress in the quantification of state factors is being made, albeit slowly, and it seems that much greater insight can be gained by pursuing the systems analogy with respect to parts of the soil system, rather than in its entirety.

CHRONOSEQUENCES

Much recent work on developing soil chronosequences has shown the way in which the nature of some geomorphic surfaces, such as terraces and erosion surfaces, has provided a boost to the understanding of soil processes. Soil chronosequences and problems with their development have been reviewed by Stevens and Walker (1970), Vreeken (1975), Yaalon (1975), and Bockheim (1980). Bockheim also developed mathematical solutions of how soil properties vary with time from 32 chronosequences described in the literature. Such studies have shown that the developmental direction or progressive pathway is not the only pathway of pedogenesis. Interruptions to pedogenesis or even reversals occur, often a result of landform instability or geomorphic processes. The rationale of chronosequences is that the time factor is the variable under consideration and other soil forming factors are considered constant. A chronosequence can be defined as a genetically related suite of soils, in which vegetation, topography and climate are similar (Harden, 1982). But, when a soil property is plotted against soil age, changes with age may or may not signify the result of a single pedogenic process acting on soils through time. Processes or changes in conditions in the history of a soil are not always recorded or preserved in morphology.

As noted above, not only may pedogenesis be interrupted, it may be reversed. This is the basis for the model developed by Johnson and Watson-Stegner (1987). Thus:

$$s = f(P, R)$$

where s represents the soil, P represents progressive pedogenesis and R regressive pedogenesis. Progressive pedogenesis, or soil progression, includes those processes, factors and conditions that promote differentiated profiles leading to physico-chemical stability. This implies the development of horizons, developmental upbuilding and/or soil deepening. It also implies increasing balance with topography and geomorphic processes, stable landsurfaces and significant correlations with landform parameters. Regressive pedogenesis or soil regression includes those processes, factors and conditions that promote simplified profiles leading to physico-chemical instability, rejuvenation processes and surface removals and retardant (non-assimilative) upbuilding. Retardant upbuilding occurs when allochthonous surface additions of aeolian and slope-derived materials impede or retard horizon differentiation or deepening. It leads to a lack of correlation with landform parameters. The distinction between the two will be a recurrent theme in this book.

Johnson and Watson-Stegner (1987) conclude that the model of soil evolution is based on observations that:

> soils are complex open process and response systems. As such they continuously adjust by varying degrees, scales, and rates to constantly changing energy and mass fluxes, thermodynamic gradients, and other environmental conditions, to thickness changes, and to internally evolved accessions and threshold conditions. Consistent with these facts is the notion embodied in the model that disturbance and change is a natural, predictable consequence of all soil and slope evolving processes (p. 363).

This acknowledges the close relationship between geomorphology and pedology.

This relationship is inherent in the four types of chronosequences differentiated by Vreeken (1975). A **post-incisive chronosequence** implies that each soil began to form in sequence but at successive times. Soils developed on a stepped river terrace sequence would fit this model. It is assumed that younger soils progress to other soils in sequence. A **pre-incisive sequence** would involve soils that began forming simultaneously but that were buried at successive, more recent, times. One of them may still be at the surface. Soils developing on a fresh glacial till surface that has then been gradually buried, would lead to such a sequence. Soils that both began forming and were buried at different times are called fully time-transgressive chronosequences. But soils from such sequences may or may not have co-existed on the surface. Thus Vreeken (1975) differentiates **time transgressive sequences with historical overlap** from **time transgressive sequences without historical overlap**. In chronosequences with historical overlap, because of erosion and deposition, a sequence of both buried and relict soils exists. Because most natural processes are space- and time-transgressive the majority of buried soil-landscapes have originated in this way. Chronosequences without historical overlap are represented by a vertical sequence of soil-landscapes such as are found between successive depositional units. These sequences can be very valuable (Chapter 12) and if each of them was understood in its three-dimensional variability and genesis, fundamental insights into the succession of pedogenetic regimes throughout geologic time would be provided.

The rate of geomorphological processes will determine the extent to which soil sequences are pre-incisive or post-incisive in character. Many standard assessments of soil-landscape relationships will also be affected by the nature of the chronosequence. As Vreeken points out the study of catenas and of other functionally integrated units of the landscape may be hampered by their internal age differentiation. Differences between soils in catenas are similar to soil differences in a post-incisive chronosequence. Lateral age differentiation along slopes in integrated drainage basins may be the norm (Chapter 4).

SOIL AND LANDSCAPE PATTERNS

The integration of pedology and geomorphology was given a conceptual boost by the formalization of the concept of the catena by Milne in the 1930s and 1940s. In the 1950s and 1960s workers such as Ruhe (1956, 1960, 1962), Walker (1962a, 1966), Butler (1958, 1959, 1967), Richmond (1962) and Morrison (1964, 1965, 1967), and many others, were advancing the integration of the two subjects. Some major research projects at this time were taking a soil geomorphological approach such as the Desert Soil Project, in the south west USA (Chapter 10) and the series of studies on the Atlantic coastal plains in the northeast United States of America (Chapter 5).

These studies demonstrated vividly that soil patterns and landscape elements often coincide and that a knowledge of one allows predictions to be made of the other. This coincidence of spatial distribution can occur on any scale. The major landform units of the 130 000 km^2 of the Llanos Orientales, the eastern tropical savanna plains of Colombia, are a case in point (Goosen, 1972). Three major units occur: the alluvial overflow plains, the aeolian plains and the high plains, each with their characteristic assemblage of soils. Each of these major landscape units is composed of individual landforms which, because of varying morphology and surface materials, subtly affect the soils. Thus, the aeolian plains possess five soil associations occupying areas with distinct physiographic features. One association comprises excessively to moderately well-drained, coarse-textured soils related to widely spaced, very shallow and often broad, drainage channels called esteros. The esteros are permanently wet and the surface horizon of the soils is characterized by a high organic matter content. A further soil association is developing as a consequence of the rejuvenating drainage system in the aeolian plain. The slightly convex relief and rather open grass vegetation, which leaves bare about 50% of the surface, results in moderate but locally severe sheet erosion.

Soil acts as the buffer zone between atmospheric and surface processes and the underlying rock. Therefore the soil profile should reflect the history of the landscape if only the signals can be deciphered. It is not surprising that Tricart and Cailleux (1972) state that the most important law of pedologic geomorphology is that chemical erosion is approximately proportional to the intensity of the soil-forming process and that the normal evolution of soils is all the more advanced as mechanical erosion is restricted. The existence of a typical, complete soil with defined horizons shows that mechanical erosion is lower than soil-forming processes. Thus, models developed to explain and predict slope evolution must include soil as a major factor. As Ritter (1988) has stressed, the evolution of landscapes is the history of regional slope development.

SOILS AND SLOPE PROCESSES

Many slope processes, such as overland flow and rill development, show extreme spatial and temporal variability making detailed process measurements very difficult. This is why many such processes are measured by their responses. Such responses are often visible in soils and some have been described by Conacher and Dalrymple (1978). Responses to the movement of materials by overland flow in the Kimberley region of Western Australia included the formation of cohesionless lenses, specific micro-roughness and near-surface cohesion of soil materials (Pilgrim, 1972). On granite slopes, cohesionless lenses up to one centimetre thick were created by the movement of material by overland flow. Mapping of these lenses on different parts of the slope enabled generalizations to be made about surface processes. Areas subject to the removal of material had lenses covering less than 10% of the land surface, whereas areas where redeposition was more significant had a 40–90% cover of lenses. The results also showed that the mean micro-roughness of mobilization/translocation surfaces was greater than that of redeposition zones. Penetration resistance was also greater on zones of translocation.

Subsurface soil water movement can also be inferred from soil properties. In podzols, these properties include repetitive variations in the numbers, size, shape, colour and pattern of mottling in the Ea horizon. Three-dimensional variations in the thickness of the Ea horizon and associated features of the illuvial B horizon are also significant (Conacher and Dalrymple, 1977). Responses on adjacent non-podzol soils were different. In gleyed-lessive soils, grey mottles elongated downslope were an indication of throughflow. Alternatively detailed measurement of processes, such as throughflow, require more detailed observations and descriptions of the morphological properties of soils such as structural peds, mottles, cutans, and intra- and interped pores. In this respect the work of Sleeman (1963) and Lafeber (1965, 1966) is important. Many developments have been concerned with the measurement of volume shapes and surface shapes of soil aggregates. Thus 'what is now very much a pedological study of a morphological property of soils . . . has evolved from the need to quantify catenary relationships of soil structures' (Conacher and Dalrymple, 1977, p. 136).

CONCLUSIONS

In a comprehensive review Birkeland (1990) has suggested that soil geomorphological studies fall within one of four areas. These are:

1. To develop a soil chronosequence framework that can be used to estimate the ages of surficial deposits.
2. To use soils as indicators of long or short-term stability. To do this it is necessary to know how long it takes to form key properties in different environments.
3. To determine soil property relations that indicate climatic change.
4. Interaction of soil development, rainfall infiltration and runoff, and erosion of hillslopes.

These themes have already been touched upon in a general way in this chapter. To a large extent the structure and contents of this book are directed towards these four main themes. Treatment of the themes is sometimes explicit but more often implicit, but however they are examined they are fundamental to an appreciation of soil geomorphology.

2

Soil landscape systems

As seen in the previous chapter, soil formation is the result of the inter-action of many processes, both geomorphological and pedological. These processes exhibit marked temporal variability, thus the soil body must be treated as a dynamic medium. Ruellan (1971) has made the distinction between those workers who attach great importance to the geomorphological processes of erosion and deposition (the allochthon-ists) and those who attribute the major characteristics of soils to pedo-logical processes (the autochthonists). But, as stressed above, soils are the result of the interaction of both sets of processes and the most realistic approach is to treat soils and landscapes as open systems and to utilize the concepts that have evolved with system analysis.

SOILS AND LANDSCAPES AS OPEN SYSTEMS

Soils and landscapes behave as open systems in that they lose and receive material and energy at their boundaries. Soils are continuously adjusting by variable degrees, scales and rates to variable energy and mass fluxes, thermodynamic gradients, and other changing exogenous environmental conditions (Johnson and Watson-Stegner, 1987). As noted in the previous chapter, a great number of theories and models of soil genesis have been advanced and Johnson and Watson-Stegner (1987) have argued that they fall into one of three general approaches: the functional-factorial approach (Dokuchaev, 1898; Jenny, 1941); the systems-process flux approach (Simonson, 1959, 1978; Yaalon, 1971) or a synthesis of both (Crocker, 1952; Jenny, 1961a; Johnson et al., 1987; Runge, 1973; Stephens, 1947). All imply that soils function as open sys-tems. This has implications both for the theoretical consideration of soil and landscape behaviour and for the choice of parameters to measure in the field in order to specify the system state. Analysis of soils and landscapes as open systems directs attention to the basic concepts involved in such a framework. These concepts have been enumerated by Strahler and Strahler (1973) in the following terms:

1. Systems possess boundaries, either real or arbitrary.
2. Systems possess inputs and outputs of energy and matter crossing these system boundaries.
3. Systems possess pathways of energy transport and transformation associated with matter within the system.
4. Within systems matter may be transported from place to place or have its physical properties transformed by chemical reaction or change of state.
5. Open systems tend to attain a dynamic equilibrium or steady state in which rate of input of energy and matter equals rate of output of energy and matter, while storage of energy and matter remains constant.
6. When the input or output rates of an open system change, the system tends to achieve a new dynamic equilibrium. The period of change leading to the establishment of the new equilibrium state is a transient state and the period of time involved will depend on the sensitivity of the system.
7. The amount of storage of energy and matter increases (decreases) when the rate of energy and material flow through the system increases (decreases).
8. The greater the storage capacity within the system for a given input, the less is the sensitivity of the system.

Not everyone would agree that dynamic equilibrium and steady state are exactly interchangeable terms, but the framework is a good one for assessing the relationships between, and within, soils and between soils and other factors. Although early models of pedogenesis were not stated purely in systems terminology, many of the concepts outlined above were tacitly acknowledged. As discussed in Chapter 1 this was also partly true of geomorphology and models of landform development. This is especially true of points 1. and 2. which, in a general way, were embodied in attempts to define soil formation in terms of state factors and state factor equations.

TYPES OF SYSTEMS

Systems have been classified in a number of ways. Chorley and Kennedy (1971) distinguished between morphological and cascade systems. Morphological systems are the formal instantaneous physical properties which are integrated to form a recognizable operational part of reality. Cascade systems are composed of a chain of subsystems which are dynamically linked by flows of mass or energy. In soils, the equivalent may be Kubiena's (1938) distinction between the soil skeleton and plasma. The skeleton is composed of the relatively stable and not readily translocated mineral grains and resistant organic bodies larger than colloidal size. The plasma is that part of the soil capable of being moved, reorganized and concentrated within the soil. It is the active part and includes all material, mineral or organic, of colloidal size and the relatively soluble material which is not bound up in the skeleton (Brewer and Sleeman, 1960). On a larger scale, hillslope hydrology integrates the morphological (slope shape, slope length, gradient angle, soil depth etc.) with the cascade components (water and sediment movement). This integration of the morphological and cascade elements produces process-response systems which exist at all spatial scales. It is the identification and analysis of such geomorphological and pedological process-response systems that is the basis of soil geomorphology.

The workings of soil systems can be treated at different levels of detail. At the 'black box' level, the whole system is regarded as a unit with no consideration of internal structure. An example of this approach would entail the measurement of precipitation as input and water emerging at the soil base as output with no consideration of pathways, stores or lags. At the 'grey box' level, a partial view of the system is adopted. At this level the soil body is recognized both as a potential regulator of water movement and as having a storage capacity. The most realistic and, therefore, most complex treatment is the 'white box' procedure where an attempt is made to identify and analyse as many of the regulators, stores and flows as possible. Water movement in individual soil layers would be assumed at this level of analysis. A mathematical simulation for soil profile development on initially undifferentiated till by Kline (1973) is an example of a 'white box' approach. The soil column is viewed as a series of compartments through which there is continuous movement of materials. Material is exported from the system by vegetation uptake, erosion and drainage and is imported through atmospheric inputs, from vegetation and animal activity. The horizontally adjacent soil compartments in this model represent transfers along pathways rather than transfers in space, whilst vertical arrangements of compartments imply transfer of material in space. The output of this model would be a series of curves, one for each compartment, showing compartmental content of the materials as a function of time.

ENERGY STATUS OF SOIL SYSTEMS

Three components can be recognized in the energy status of soil systems. There is a decay component in which the energy status gradually declines and eventually the system should be brought to a state of virtual exhaustion. Soils existing on very level ancient erosion surfaces approach this condition. Lack of relief means that potential energy is at a minimum and vertical movement of soil water is the only possibility. Thick soils and possibly thick weathered regolith also inhibit chemical action at the interface or weathering front between regolith and rock. These soils will still have a cyclic or rhythmic component imposed by diurnal and seasonal weather and climatic patterns. There will also be a random component in which input of energy and matter occur irregularly, such as rainstorms. The soil systems evolve in response to all these components. Energy can flow through soil systems in a variety of ways and one of the problems is how to assess the energy status of a soil body. This has been attempted in an enterprising fashion by Runge (1973).

One of the major drawbacks in Jenny's synthesis is that the state factors are discrete, non-overlapping units and offer little opportunity of obtaining the data necessary to determine the differential rates of change which are an essential feature of soil formation. Runge (1973) has argued that some vectors are more important than others in controlling soil development. In his model, the energy source is the gravitational or potential energy that is available to the soil system when water runs off the soil surface or percolates through the soil profile. Soil development, at a particular site, is dependent on the relative amount of water running off, and therefore not contributing to soil development, versus the amount infiltrating and therefore available to influence soil formation. This is why the geomorphology of the area is so important because if the change in soil development between soil profiles is being examined, it is essential to have the same soil material and equal stability within the landscape. Landscapes and soil profiles become a record of how the internal and external energy fluxes have been dissipated over time.

In the energy model, the analogy of systems and thermodynamics is used. The first law of thermodynamics states that the total amount of energy remains constant. The second law states that, with time, systems develop towards states of maximum entropy. Entropy is an expression of the degree to which energy has become unable to perform work. The state of lowest available energy and maximum disorder occurs when entropy is at a maximum, and to decrease entropy and increase order energy has to be imported from another source. Runge (1973) argues that loess soil parent material is at maximum disorder or has no profile development, alternatively, a well-developed soil profile with horizon differentiation is considerably more ordered. This leads to the idea that water flow through the soil profile is the principal source for increasing order and decreasing entropy of the soil body.

The essence of the model developed by Runge (1973) and Smeck and Runge (1971 a, b) is that soil development is a factor of organic matter production, the amount of water available for leaching and time. In the model phosphorus was used as a surrogate for organic matter production because, under natural conditions, it is only supplied by the soil parent material. Phosphorus is also considered to be essentially immobile in soils but, during the time spans involved in soil formation, redistribution does occur. As pH drops, the relatively soluble forms of phosphorus decrease and more occluded forms increase. Therefore, the relative amounts of the different phosphorus forms may serve as a measure of soil development.

SOIL-LANDSCAPE SYSTEMS

Dijkerman (1974) has stressed that scientific explanation is a satisfactory answer to a why or how question. The first question seeks a genetic explanation to questions such as how did the system originate and develop. An explanation to a question of this nature must be in terms of the sequence of events that produced the situation being explained. The second question seeks a functional explanation to questions such as how does the system function. It asks for an assessment of the status and role of the many forces and factors acting on the system. In answering these questions with respect to soils it is necessary to consider the significance of topography and position. The geomorphological history of an area is fundamental for an answer to the first question, and an assessment of the interactions of geomorphological and pedological processes is important for a satisfactory answer to the second question. Soils do not exist in isolation but are organized within the landscape. This is embodied in the concept of pedons and polypedons (Simonson, 1968). A pedon consists of a small volume of soil starting at the surface and extending downwards to include the full set of horizons. It must be large enough to include a full set of horizons and to permit observation of the boundaries between them. Larger units composed of a number of similar contiguous pedons are called polypedons. The relationships between these are depicted in Figure 2.1.

The interaction between soils and topography or pedology and geomorphology can be treated at several levels. In the state factor model the interactions can be construed as a function of the topofunction and lithofunction being acted upon by climatic and biotic factors at some point in time. This approach does little to assess the detailed operations of the processes; it is a black box approach. More realistically, soils and landform systems must take account of the flux of materials and energy through the systems and this depends not only on topography or slope angle but relative position. This involves consideration of balance between input and output, i.e. equilibrium.

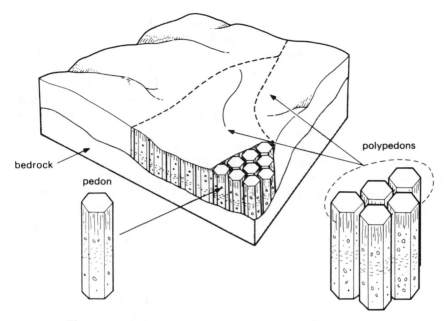

Figure 2.1 Relationships between pedons and polypedons.

DENUDATIONAL BALANCE

The thickness of soil and regolith at any point will depend on the relative rates of soil removal and soil formation. At some sites removal will be minimal and deep soils and regolith will develop, whilst at other, more erosionally active sites, soils will be kept thin and permanently youthful. These aspects have been embodied in the distinction between accumulative and non-accumulative soils (Nikiforoff, 1949). In geomorphology, the situation has been conceptualized by Jahn (1968) in terms of denudational balance. The three components that Jahn used were the accumulation of material by *in situ* production of waste and by inflow of material from upslope, and the removal of material by slopewash, surface deflation and mass movements. The arrangement of these factors yields the following three alternatives:

$$A = S + M \quad A < S + M \quad A > S + M$$

where A = the accumulation of slope material, S = the processes of slopewash and surface deflation and M = mass movement. Soil thickness will, thus, remain constant or increase and decrease according to the efficacy of the respective processes. If transport processes are more rapid than weathering, only a thin soil cover will exist because material is removed as fast as it becomes loose. The development of such a site is then said to be weathering-limited. If weathering rates are more rapid than transport processes, a thick soil cover develops and the site is said to be transport-limited. Hillslope and soil development, in weathering-limited situations, depend on the variations in weathering rate and the rate of transport is reduced to the rate at which fresh material weathers. On transport-limited sites, soil and slope development depend on the transporting capacity of the processes and the rate of weathering is reduced to an equilibrium value less than its potential maximum by the increase in soil thickness. In thin soil, very little water is retained and weathering rates are low. In very thick soils, water moves so slowly towards the weathering front that the rate of weathering is again below the potential maximum. Thus, weathering and soil formation are at a maximum at intermediate soil thicknesses. In practice, the rate of soil formation will fluctuate around a mean value as the relative efficiency of transporting processes varies. Carson and Kirkby (1972) have argued that these different controls lead to different slopes and sequences of slope development. Weathering-limited slopes, possessing thin soils, have prominent straight sections with important threshold angles and develop by parallel retreat. Transport-limited slopes, possessing thick soils, are essentially convex-concave and become progressively less steep with time. This is a very specific example of the interaction between soils and long-term landscape development.

INSTABILITY IN SOIL GEOMORPHIC SYSTEMS

Transitions between the various types of denudational balance often occur quite rapidly during phases of instability. Many earth scientists now believe that change in the landscape takes place rapidly over a short-time period and that short periods of change are separated by longer periods of comparative stability. Such ideas have been embodied by Butler (1959) in his K-cycle concept (Chapter 12). Each cycle will have an unstable phase (ku) of erosion and deposition followed by a stable phase (ks) accompanied by soil development.

One of the main characteristics of landscape systems is their fluctuating stability. Concepts proposed by Gigon (1983) are useful in examining landscape stability. Constancy of system occurs where disturbance factors external to the system are absent and little or no oscillation of system parameters occurs. Large oscillations produce cyclicity. If external disturbance factors occur but little or no system oscillation results, the system exhibits resistance. Large oscillations indicate resilience or elasticity. Some landscape systems are more prone to external disturbance factors than others. Soil-geomorphic systems in mountains are probably in this category (Gerrard, 1990a). Dynamism in soil-geomorphic mountain systems results from combinations of external influences such as tectonic activity, rejuvenation and climate fluctuations and internal changes such as weathering, mass movement and land use changes. The dynamic nature of mountains encourages recurrent stability over the long term in relatively immature systems which would otherwise become mature or even post-mature.

The mountains of New Zealand provide some of the most dynamic geomorphic systems to be found anywhere. A high relief, tectonically active youthful landscape, combined with high precipitation amounts, produces high erosion rates and dynamic landscapes (O'Loughlin, 1969; Mosley, 1978; Grant, 1981; Tonkin et al., 1981; Griffiths, 1981). O'Connor (1980, 1984) has examined some New Zealand mountain systems using Gigon's stability types. The high-rainfall regions of Westland and Northwest Canterbury, with periodic rock avalanches, debris avalanches and debris slides, are interpreted as demonstrating cyclic stability or cyclicity. The drier mountain system of Central Otago appears to exhibit some cyclicity of periglacial activity caused by climatic oscillations but superimposed on a general constancy. The Central Canterbury fold mountains were classed as possessing either cyclic stability or natural endogenous instability. Endogenous instability may have been present even with apparently mature soils, and before the advent of man some soil-geomorphic systems show a tendency towards irreversible change. O'Connor (1984) concludes that regional differences in stability of the landscape reflect differences in both the frequency and pace of soil-biotic succession.

TWO-DIMENSIONAL SOIL-LANDSCAPE SYSTEMS

One of the simplest ways of analyzing soils in the landscape is to assume they are two-dimensional bodies existing at some point on a topographic transect. The topographic transect is usually, but not necessarily, a valley-side slope profile. It is embodied in the concept of the catena and numerous examples of this approach can be cited, some of which are analyzed in the next chapter. Empirical approaches to fitting topofunctions to soils have been summarized by Yaalon (1971). In terms of the processes involved it may be necessary to subdivide this 'line' system into separate working subsystems.

Just as the inner workings of the system can be assessed at different levels of complexity so can the input, throughout and output. This can be illustrated by the movement of water through the upper soil horizons on individual slopes. In the simplest case, it can be assumed that all water enters the soil system at the top of the slope and leaves it at the slope base (Figure 2.2(a)). This situation is extremely rare but may be convenient for simple analysis. Slightly more realistic is the situation where inputs and outputs are still *en bloc* but the soil system has been

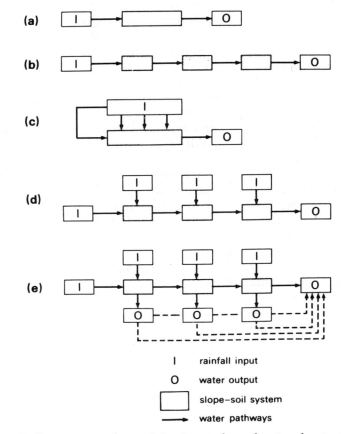

Figure 2.2 Different ways of organizing input, throughput and output in systems (from Chorley and Kennedy, 1971).

subdivided and transfers between these subelements considered (Figure 2.2(b)). Using the soil water movement analogy, this is equivalent to subdividing soils on the basis of slope position and form, e.g. crest, mainslope and footslope. The third case (Figure 2.2(c)) allows input to occur throughout the length of the slope but the slope is not subdivided. A more complex case is where inputs and throughputs are seen to be composed of discrete, though related, components on different slope zones (Figure 2.2(d)). The most realistic approach is where inputs, throughputs and outputs are all composed of discrete units (Figure 2.2(e)). In the analogy being used, this allows percolation to lower soil horizons as well as throughflow, deep percolation and groundwater flow. Although formulated differently this is essentially the conceptual embodiment of the catena.

The geochemical landscapes and geochemical soil sequences of Glazovskaya (1968) involve similar concepts. Landscapes and soils that are adjacent, but at different elevations, are united by the later migration of chemical elements into a single geochemical landscape. Two examples, one from the central Tien Shan in central Asia and the other from Norway, illustrate the principles involved. In the central Tien Shan, granite surfaces outcrop at the tops of slopes with the lower slopes

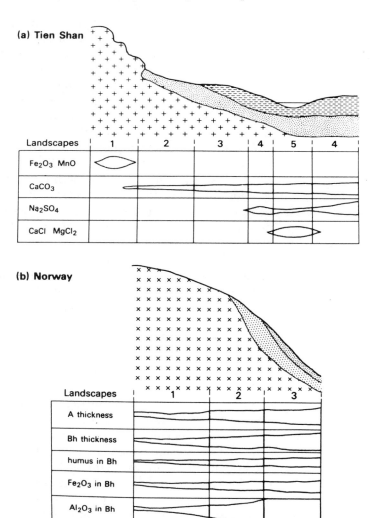

Figure 2.3 Geochemical soil sequences of: (a) Tien Shan and (b) Norway (from Glazovskaya, 1968).

composed of reworked weathered granite debris (Figure 2.3(a)). Zone 1 is an eluvial landscape of granite rocks with desert varnish, while zone 2 is a transeluvial landscape of denudation surfaces with slightly carbonaceous polygonal soils. Zone 3 is mostly an accumulative landscape of morainic hillslopes with highly carbonaceous takyr-like desert soils. Zones 4 and 5 are less well-drained with meadow saline soils giving way to landlocked basins with wet solonchaks. The more readily soluble salts such as calcium and magnesium chlorides reach the lowest levels. Sodium sulphates are partly retained on the lower slopes. Although soils in the Norwegian sequence are different the principles are the same (Figure 2.3(b)). Due to solifluction the depth of soil increases towards the lower parts of the slopes and the thickness of both A and Bh horizons also increases. Chemical status changes as there is an intersoil migration of aluminium fulvates with subsequent accumulation in the lower part of the slope. There is a close link between this analysis and the energy models of Smeck and Runge (1971a, b) described earlier.

The nine-unit landsurface model of Dalrymple *et al.* (1968) is essentially a two-dimensional approach although it can be extended to encompass entire drainage basins. It is a model based both on form and contemporary geomorphological and pedogenetic processes. It attempts to subdivide slope profiles and yet at the same time to integrate the components by considering material and water flow. The model is depicted in Figure 2.4. On units 1 and 2, pedogenetic processes and vertical water movement dominate. The convex creep slope (unit 3) is characterized by both pedogenetic and geomorphological processes. Units 4 and 5, fall face and transportational midslope, are controlled by the processes of weathering and mass movement. The colluvial footslope (unit 6) contains both geomorphological and pedogenetic processes. The alluvial toeslope (unit 7) is controlled by subsurface water movement and periodic incursions by the river in flood. Units 8 and 9 are fluvially controlled. Since the model was presented in 1968 it has been refined, culminating in a seminal publication in 1977.

The various units or components are separate process-response systems in the sense used earlier. In some cases the integrative factor is the mobilization, translocation and redeposition of materials by overland flow. On other units, the subsurface movement of water is the more important. Thus, unit 2 is defined as an area where the response to mechanical and chemical eluviation by downslope subsurface soil-water movements distinguishes this from other parts of the landscape and unit 5 is defined by the response to transportation of a large amount of soil material relative to other units (Conacher and Dalrymple, 1978). These process-response pedogeomorphic units can be mapped both in detail and over extensive landsurfaces.

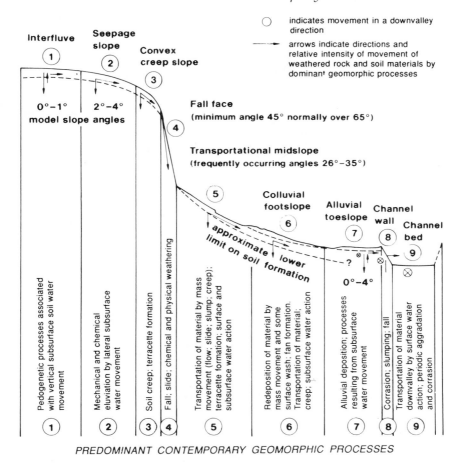

Figure 2.4 Hypothetical nine-unit landsurface model (from Dalrymple *et al.*, 1968).

The nine-unit landsurface model has been very influential in furthering the integration of pedology and geomorphology. It is an extension of the catena concept, discussed in Chapter 3. The units can also be used to examine spatial relations within drainage basins (Chapter 4). However, it may not be applicable to all landscapes. It seems to be most applicable to 'mature' relatively stable temperate landscapes. There are a number of problems in applying the model to highland and mountain landscapes (Fowler, 1990). In such environments other models, such as those of Ryder (1981) and Caine (1974) might be more appropriate. Both models are essentially landform models but within which it is relatively easy to add the soil component. Other soil-geomorphic models have been proposed specifically for mountain regions (e.g. Richmond, 1962;

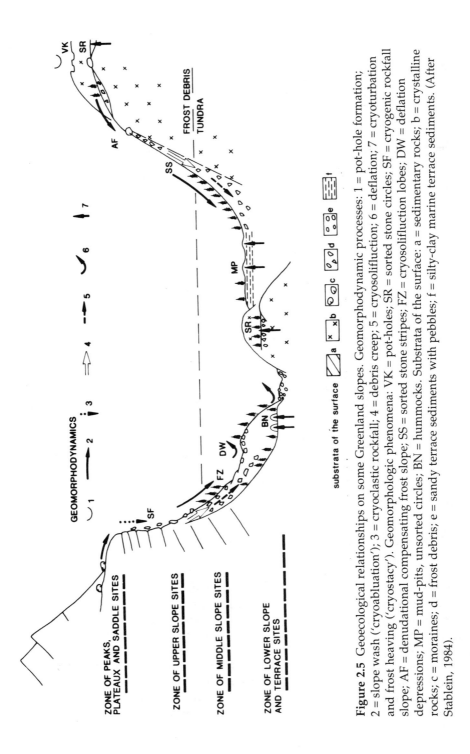

Figure 2.5 Geoecological relationships on some Greenland slopes. Geomorphodynamic processes: 1 = pot-hole formation; 2 = slope wash ('cryoabluation'); 3 = cryoclastic rockfall; 4 = debris creep; 5 = cryosolifluction; 6 = deflation; 7 = cryoturbation and frost heaving ('cryostacy'). Geomorphologic phenomena: VK = pot-holes; SR = sorted stone circles; SF = cryogenic rockfall slope; AF = denudational compensating frost slope; SS = sorted stone stripes; FZ = cryosolifluction lobes; DW = deflation depressions; MP = mud-pits, unsorted circles; BN = hummocks. Substrata of the surface: a = sedimentary rocks; b = crystalline rocks; c = moraines; d = frost debris; e = sandy terrace sediments with pebbles; f = silty-clay marine terrace sediments. (After Stäblein, 1984).

Birkeland, 1967; Parsons, 1978; Tonkin *et al.*, 1981). Stablein (1984) has produced a dynamic catena profile applicable to arctic and alpine areas (Figure 2.5). He defines four valley-side zones:

1. A belt of peaks, plateaux and saddle sites on which frost weathering, solution weathering, slope wash, cryoturbation and wind deflation are common.
2. An upper-slope belt where frost weathering, rockfall and debris creep occur.
3. A mid-slope belt on which slope wash, solifluction, nivation and slope dissection are active.
4. A lower belt, often with terraces, where cryoturbation, frost heaving and wind action occur.

Variations in soil properties explain the dynamics of the system (Stablein, 1977, 1979). In the zone of peaks and plateaux, bedrock surfaces are pitted by weathering pans, solution hollows and tafoni. More sheltered zones encourage the accumulation of frost-weathered debris and there is sufficient material and moisture for freeze–thaw cycles to produce patterned ground. Slope wash is active on upper slopes with little vegetation. Rill and gully formation is common on slopes steeper than 25°.

A more overtly pedological model for mountain slopes is the synthetic alpine slope model of Burns and Tonkin (1982). The model, developed for the southern Rocky Mountains, is based on the K-cycle concept and divides the alpine tundra zone into three geomorphic provinces (Table 2.1). The provinces are:

Table 2.1 Soil-geomorphic relationships in the Alpine tundra zone (after Burns and Tonkin, 1982)

Alpine province	Dominant geomorphic process	Soil variability control by state factors	K-cycle stable phase, period of duration*
Ridge-top	periglacial	spatial > temporal	long/medium
Valley-side	gravity some glacial	temporal > spatial	short
Valley-bottom	glacial fluvial gravity	temporal > spatial	short/medium

* 'Long': greater than 15 000 years; 'medium': 5000 to 15 000 years; 'short': less than 5000 years.

1. The ridge-top tundra province. This consists of broad interfluves that have not been glaciated in the late Pleistocene and can be interpreted as a persistent zone characterized by relative stability. This stability is reflected in mature soils with deep profiles.

2. The valley-side tundra province. This is located on the steep slopes
   of valley sides. Such slopes have been undergoing dynamic
   change throughout the last 10 000 years and soils reflect this instability,
   being predominantly young, thin soils intermixed with rock outcrops.
3. The valley-bottom tundra province. This is primarily present in cir-
   que floors but can be present further downvalley. Soils have developed on a variety of glacial, fluvial and colluvial materials.

The 'maturity' of soils reflects the lengths of the various stable phases:
short phases (less than 5000 years) are estimated from glacial soils that
have not reached steady state characteristics; medium-length phases
(5000–15 000 years) are characterized by soils that have minimum
steady-state requirements; and long periods (greater than 15 000 years)
are based on the ages of well-developed glacial soils.

   Detailed soil-geomorphic relations are represented in subdivisions of
each province. This can be seen in an examination of the ridge-top province (Figure 2.6). Soil terminology follows that of Soil Survey Staff (1975):

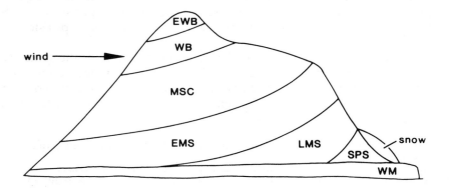

Figure 2.6 Schematic representation of the synthetic alpine model (after
Burns and Tonkin, 1982) **Note**: abbreviations are explained in the text.

1. The extremely windblown (EWB) sites are located on the crests of the
   drainage divides. Soils at such sites (90% Dystic Cryochrept, 10%
   Typic Cryumbrept) are poorly developed and well-drained with thin
   sandy A horizons over thin cambic B horizons.
2. The windblown (WB) sites are found from the peaks to 30% downslope. WB soils (80% Dystic Cryochrept, 20% Typic Cryumbrept) are
   similar to EWB soils but with thicker cambic B horizons and patches
   of loess. Loess occurs in some A horizons.
3. Minimal soil cover (MSC) sites occur in cols and on plateaux. MSC
   soils (80% Pergelic Cryumbrept, 20% Dystic Pergelic Cryochrept)

possess thick, fine-textured A horizons of aeolian origin overlying cambic B horizons. They tend to be some of the best-developed soils. A horizons contain the highest organic matter content.

4. Early melting snowbank (EMS) sites are found on middle to lower, often gentle, slopes. EMS soils (60% Typic Cryumbrept, 30% Pachic Cryumbrept, 10% Dystic Cryochrept) are comparatively well drained with the thickest A horizons.

5. Late melting snowbank (LMS) sites are generally found on lower slopes usually in leeward nivation hollows. Sites are generally rocky with sparse vegetation. LMS soils (100% Dystic Cryochrept) are poorly developed, moderately well-drained soils overlying weakly-developed cambic B horizons.

6. Semi-permanent snowbank (SPS) sites are found in nivation hollows with little vegetation cover. SPS soils (headwall, Lithic Cryothent; nivation hollow, Pergelic Cryoboralf or Pergelic Cryochrept) are poorly developed.

7. Wet meadow (WM) sites occur below snowbank sites in depressions and on turf-banked terraces and lobes at the base of slopes. They are characterized by bog vegetation. WM soils are poorly drained with either A and/or O horizons of variable thickness overlying gleyed and mottled B and C horizons.

THREE-DIMENSIONAL SOIL-LANDSCAPE SYSTEMS

The necessity of viewing the soil as a component of the broader landscape has always been recognized but the means of formulating this has been difficult. Consistent surface geometric soil patterns in a variety of scales which are approximately coincident with topographic patterns demonstrate the spatial relations between soils and landforms. The major geometric forms that these relationships can take have been classified by Fridland (1974) and are:

1. Dendritic, linear-dendritic and streaming forms, connected with various forms of erosional relief such as river valleys.

2. Rounded, spotted (including ring-like) forms, connected with various depressional forms (sink holes, former lakes) and hilly morainic relief.

3. Linear and wavy-linear forms connected with different linear forms of an accumulative origin.

4. Streak-phacoid forms characteristic of modern and ancient floodplains and deltas.

5. Fan-shaped forms characteristic of alluvial fans and deluvial trains.

These patterns are the result of the interactions of many processes.

One of the integrating themes is the relationship between water flow, soils and topography. Relationships have been established between throughflow and slope profile curvature (e.g. Anderson and Burt, 1978b; Speight, 1980; Sinai *et al.*, 1981; Hall 1983) and plan curvature (Troeh, 1964). Other workers (e.g. Lanyon and Hall, 1983) have presumed that such relationships exist. England and Holtan (1969) differentiated slopes into upland, hillside and bottomland sites, whereas Dunne (1978) adopted a more involved scheme, differentiating lower and upper parts of valley floors and shallow, moist swales, lower concave portions of well-drained hillsides and straight well-drained hillsides. Hack and Goodlett (1960) classified slopes as nose, sideslopes, hollows, footslopes and channel ways. Runoff on nose slopes or spurs was proportional to a function of the radius of curvature of the contours. On sideslopes, runoff was proportional to a linear function of slope length, and in the hollows was proportional to a power function of slope length. In the channelway, runoff was proportional to a power function of channel length whilst the footslope was transitional between sideslopes and channelway.

The three-dimensional aspects of slopes was examined by Hewlett and Nutter (1970) when they were able to establish relationships between soil thickness and channel hydrographs. But it is only with the advent of high-speed computers that the three-dimensional situation can be handled. The best example of this is the attempt by Huggett (1973, 1975) to simulate the flux of plasmic material in an idealized valley basin. A valley basin was chosen as the basic organisational unit for both the geomorphological and pedological materials and processes. The model is really an extension of the nine-unit landsurface model in that it argues that definable flowlines of material can be organized into soil-landscape system units. In a valley system, flowlines diverge and converge; convex contour patterns lead to divergent flow and concave patterns to convergent flow. Extension of this into the third dimension allows the theoretical direction taken by infiltrating water to be examined. Zaslavsky and Rogowski (1969) found that convex slopes led to divergent infiltration and concave slopes to convergent infiltration. Thus, divergent throughflow of water on spurs is enhanced by divergent vertical flow.

The computer-simulated pattern of change in the concentration of a plasmic constituent in an idealized valley is shown in Figures 2.7(a) and (b). Figure 2.7(a) shows the pattern of concentration values one depth increment below the soil surface, after 2- and 4-time steps. The accumulation of material in the hollows is evident. Figure 2.7(b) depicts the changes in concentration at the same position in the soil body along two side spurs and along the hollow thalweg. Material has moved down all the slopes, but spur 1 has lost material over most of its length whereas spur 2 has lost material in its upper convex section and gained material

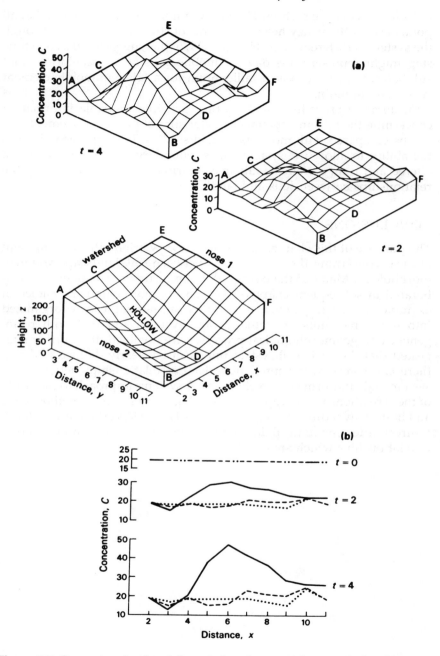

Figure 2.7 Computer-simulated flux of plasmic material in an idealized basin (from Huggett, 1975).

in its lower concave section. The thalweg line has gained material at all points except the valley head. Different constituents will move through the system at different rates. Huggett (1975) has suggested that one time step might represent one day for mobile salts such as chlorides and sulphates, but as much as a millennium for a fairly immobile element such as aluminium.

An iterative procedure, such as this, provides valuable information concerning the rate and spatial characteristics of plasma movement that can be compared with more static field studies. The approach also has the ability to allow the system states, boundary conditions, etc. to be altered and is an extremely valuable contribution to understanding the relationships between soils and landforms.

CONCLUSIONS

The purpose of this chapter has been to review briefly the development of ideas concerning the relationships between soil, landscape and geomorphology. Many of the examples given a cursory treatment are elaborated in subsequent chapters. The philosophy behind the choice of examples is that the movement, or flux of material, can be organized into functional units or soil-landscape systems and that both pedogenetic and geomorphological processes are involved. It is further suggested that the basic soil-system units can be clearly defined and that there is a complex, but intelligible, web of relationships between soil system and landform system elements. A systems approach to the study of these relationships is appealing in its logic and rigorous methodology and in the way it directs a stage by stage advance to particular problems. It directs attention to the delimitation of entities or parts, and the choice of relationships which are of interest.

3

The catena concept

FORMULATION OF THE CONCEPT

The realization that particular slope forms were associated with particular soil sequences led to the formulation of the concept of the catena. Milne originally defined a catena as 'a unit of mapping convenience . . ., a grouping of soils which while they fall wide apart in a natural system of classification on account of fundamental and morphological differences, are yet linked in their occurrence by conditions of topography and are repeated in the same relationships to each other wherever the same conditions are met with' (Milne, 1935a, p.197). Since then catenas have been recognized in a variety of areas and under a variety of climatic conditions, but the concept is one which has been subject to a great deal of discussion and controversy.

The real significance of catenas lies in the recognition of the essential processes involved in catenary differentiation and not in the formal appearance of its product. It is the interaction of soils and landforms, and therefore soil processes and geomorphic processes, which is the key to catenas and the reason why the concept has been so important in soil and landform studies. But the wide applicability of the concept is complicated by considerations of parent material variations and climatic differences. The temporal as well as spatial aspects of the soils are also important. Thus, before specific examples of catenas from different parts of the world are analyzed, these issues need clarification.

THE CATENA AND THE PROCESSES OF EROSION

Milne (1936a) was one of the first to include the processes of erosion as a major factor leading to the differentiation, under constant climatic conditions, of several different but related soils usually from a common original material. The example he used to illustrate this was a residual granite hill and associated slopes in East Africa (Figure 3.1). A shallow

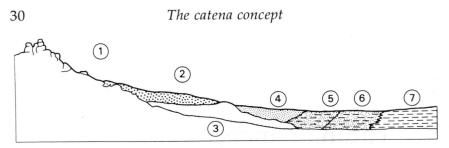

Figure 3.1 Soils of an East African catena (after Milne, 1936a).

grey loam (1) formed by weathering of the granite surfaces has worked downhill by creep and slow erosion to act, on the footslope, as the parent material on which a deeper soil (2) of the red earth group has developed. At the base of the red earth profile, where a temporary accumulation of seepage occurs in the wet season, a horizon of coarse granitic grit (3) in a black rusty ferruginous cement has formed. Occasional storm-water running over the surface has gradually pared off the topsoil and the material has travelled differentially according to particle size, so that by a cumulative effect a zone of washed sand (4) has covered the footslope, with silty or clayey sand (5,6) beyond it, and clay has accumulated on the level bottomlands (7). At all stages erosion has been slow and non-catastrophic and soils have borne their appropriate vegetation and been developing towards maturity.

Milne further suggested that the character and proportionate extent of soils varies with maturity of the topography, with underlying lithology and with new cycles of erosion. Therefore, the physiographic and geomorphic evolutions of the landscape are both involved in the catena concept. The soil profile changes from point to point in accordance with conditions of drainage and past history of the land surface, and soil differences are brought about by 'drainage conditions, differential transport of eroded material and leaching, translocation and redeposition of mobile chemical constituents' (Milne, 1936b, pp.16–17). This concept is extremely important for the way it relates the soil to the processes operating and to the past history of the landscape. It is therefore very difficult to make any useful distinction between slope genesis and pedogenesis and this means that a better understanding of the soil should be sought in a geomorphological evaluation of the soil landscape. But, equally, a better understanding of the geomorphology of a region should be sought in a study of the soil.

CATENARY DIFFERENTIATION

The differences between the soils of a catena are generally related to differences in their position and their drainage characteristics so that emphasis is placed on the difference between freely drained upper parts of slopes and imperfectly to poorly drained lower portions. This provides a continuum between those sites where the influence of soil moisture is at a minimum and those sites where maximum influence of soil moisture is felt. Slope steepness is one of the most important factors that causes a variation in soil moisture conditions as the steeper angles reduce the amount of water percolating through the soil and increase the removal, perhaps through accelerated erosion, of the upper portions of the soil profile. The essential feature is that soil and water can and do move downslope. For these processes to have their greatest effect the ground surface must slope downward continuously from the crest to the base of the slope, and it is incorrect to apply the term catena to landforms which lack this feature.

The main processes of catenary differentiation are surface wash, soil creep, solution and rapid mass movements. These processes vary in their relative importance and effectiveness with climate and slope. The importance of creep in the formation of catenas has been recognized by a number of workers (e.g. Adams and Raza, 1978; Bishop *et al.*, 1980; Furbish, 1983). Erosion by creep occurs on convex slopes and deposition on concave slopes (Nash, 1980) thus there should be some relationship between slope curvature and soil properties that are affected by creep erosion and deposition. Surface wash rates vary widely: in dry savanna areas, wash is very effective but in areas of higher rainfall the increased vegetation cover gradually reduces wash to a minimum near the savanna/rainforest boundary; at higher rainfalls the trend is apparently reversed because the controlling influence of vegetation cover is already at a maximum whereas rainfall intensity may be increasing. Although surface wash under tropical rainforest occurs, its overall importance under such conditions is still largely unknown (Rougerie, 1960). Most of Rougerie's observations were made in semi-evergreen forest and in a region where man-induced changes in vegetation have been important. In tropical forests that have been cultivated, such as French Guiana, surface wash seems to be insignificant (Cailleux, 1959). Measurements on experimental plots near Abidjan have also revealed little surface flow (Tricart, 1972). The role of soil creep is still largely unknown in tropical rainforest areas and it is probably of much greater relative importance in humid temperate regions.

Early workers in East Africa, whilst ascribing varying importance to these processes, all agreed that they were significant in soil formation. Evidence for the potential of mass movements in these environments has been provided by work on the mechanics of the deep-red clays of East Africa (Newill, 1961; Coleman *et al.*, 1964). In general, the importance of mass movement in the shaping of slopes in tropical areas has long been recognized and man-induced landscape changes can also be significant.

The downslope movement of material in solution is of great significance in catenary studies. Thus, soils are affected by the influx of soluble materials, especially bases, from higher up the slope and this leaching and redeposition of material constitutes a strong physical link between the members of a catena which is closely analogous to the link between the A and B horizons of a soil profile. This was first stressed in a classic early paper by Greene (1947). Considerable amounts may be removed from slopes in this way especially under tropical rainforest conditions.

The relative importance of wash, solution, creep and rapid mass movements depends not only on climate but also on slope angle and distance from the slope crest. The rate of erosion by surface wash increases somewhere between linearly with slope angle and as the square of the angle, and approximately in proportion to the square root of the distance from the crest. At steep angles the rapid forms of mass movement become important although the critical angle at which they become significant varies with the type of bedrock and the regolith conditions. On gentle slopes in tropical regions chemical removal is the most important process in climates such as those of savanna areas, although in humid temperate regions it may be subordinate to soil creep.

Each catena is, therefore, the result of the complex interrelationships between soil and slope processes and will be governed by the differing ratio of erosion to deposition occurring on different parts of the slope. From the pedogenetic point of view, all country which has relief consists of zones of removal, transference and accumulation, the limits of which can be peculiar to each transferable constituent or to each group of constituents of comparative mobility. This applies equally to the most simple of landscapes and to great geomorphological units. This erosion-deposition relationship can be quite complicated on individual slopes though it is usually the upper parts of slopes that lose material and the lower parts that gain it. This type of catena, where erosion on the upper and deposition on the lower slopes has caused a variation from a uniform soil cover has been termed an erosion catena by Ollier (1976). He cites the example described by Ellis (1938) from Manitoba where the lower B and C soil horizons have been gradually exposed by erosion at the top of the slope. A more complex situation has been described by Webster (1965) in Zambia where erosion by surface wash is greatest on the lower steeper parts of the slope leading to the preferential removal of fine soil particles and leaving a coarse-grained soil at the base of the slope.

Young (1972b, 1976) has distinguished between static and dynamic causes leading to catenary differentiation. Static causes are governed by site differences alone, irrespective of the position of the site, and include effects of slope angle and the depth of the water-table. Dynamic causes are brought about by the position of the site with respect to the slope;

they are mainly the downslope transport processes described above. On most slopes there is clearly an interplay between these two groups but it is conceivable that an extremely permeable parent material, or very gentle slopes, might inhibit downslope movement and then static causes will essentially control the development of the catena. These conditions are normally found on sand dunes, beach ridges or volcanic slopes such as scoria and ash cones. Conversely on steep slopes, or on very impermeable rocks, dynamic causes will be dominant, whilst on other slopes static causes may be dominant on one part and dynamic causes on another. Thus, if gentle crestal slopes exist, static causes may be reversed on the steeper slopes. This complex situation is well-shown on loess slopes in Iowa described by Huddleston and Riecken (1973). In midslope situations the soil has inherited the combined initial sorting of the loess, the downslope sorting during slope evolution and the distribution of iron carbonates produced by weathering. The summit portions have escaped the erosive processes but both the shoulder and the toe portions of the slope show signs of the simultaneous interaction of erosive and pedological processes. The ideas embedded within Penck's aufbereitung concept and the distinction made by Holmes (1955), between derivation and wash slopes, help explain how soil becomes differentiated on slopes.

AUFBEREITUNG CONCEPT

This account of Penck's aufbereitung concept relies heavily on the summary and assessment by Beckett (1968). As weathering progresses, soil becomes reduced, in the sense that there is a reduction in its average particle size. The potential mobility of the soil increases as the particle size decreases and all reduced material above a local base level is 'metastable', i.e. vulnerable to removal. The maximum rate of natural denudation is the rate at which surface material which is almost, but not quite mobile, becomes mobile. Thus, on any particular gradient there is a degree of reduction beyond which soil is too mobile to remain *in situ*. Therefore the development of the landscape will be a function of this metastability of the soil; which is a function of the ease with which soil is detached and transported. But, Penck argues, landscape development depends also on the metastability of the particular site, which is a function of slope gradient and proximity to the next major break or change of slope on the slope profile. An encroaching pediment will thus affect the soil on the upper surfaces long before that soil is affected physically by erosion. The aufbereitung concept stresses that even on relatively uniform surfaces there will be sufficient differences to affect local rates of soil development. This is also implicit in the synthesis put forward by Holmes (1955).

Penck's argument is that the landscape develops by the upslope encroachment, upon each slope element, of the one below it of lower gradient and greater degree of reduction. This would lead to a landscape of concave slopes formed by slope retreat. However, the analysis by Beckett (1968) has shown that the relationship between gradient and degree of reduction applies only to those parts of a slope on which there is no accumulation of transported material. The relation applies to the convex and linear portions but not to the concave portion. If Penck's arguments are applied only to the convex and linear portions, the landscape must develop by crest-lowering and crest-rounding and lessening of gradients. If these conclusions are realistic then one is forced into 'the ironical position that Penck's argument, when applied only within its implicit limiting conditions and for the elements of a slope to which it can be shown to apply, provides support only for the so-called Davisian cycle of landscape development' (Beckett, 1968, p.19).

DERIVATION AND WASH SLOPES

Holmes (1955) argues that the slopes of a landscape can be classified as derivation slopes or wash slopes: derivation slopes correspond to scarps and washslopes correspond to pediments. Derivation slopes are unstable and provide a source of sediment which is deposited on, or transported across, the wash slopes. Also, gross derivation or wash slopes are potentially unstable areas, and the proportion of derivation slopes determines the rate of denudation. The difference between derivation and wash slopes is not wholly dependent on slope gradient but also depends on local factors, such as vegetation and climate. This implies that landscapes do not necessarily develop by continuous downwasting or backwasting, but that development will take place when slopes are unstable and cease when they become stable.

These ideas imply that the landscape can be divided into three major zones: zones where material is being eroded, zones where material is being deposited and, perhaps, zones which are neither losing nor gaining material. The first two are equivalent to the 'sloughing' and 'accreting' zones of Butler (1959) (Chapter 12). Continuous soil development will only take place on the third zone. The accreting zone will show sequences of buried soils while soils on sloughing zones might be kept perpetually youthful. But, the location of these zones on slopes will vary with time, thus sloughing zones will become stable and accreting zones may become unstable. The geomorphological end-product will be a relatively smooth erosion surface and it will only be pedological information that will show how the surface was developed.

SOIL CHANGES WITHIN CATENAS

The result of the processes described above is to produce a series of changes in soil properties from the upper to lower members of the catena. The variation in soil colour is one of the more obvious sequences and is typical of many West African catenas. Upland, well-drained soils are usually reddish-brown, the colour showing the presence of non-hydrated iron oxide in the soil. The iron is well dispersed and usually partly attached to the clay fraction, thus the clay itself appears red. On middle and lower parts of the slope, drainage is slower, partly because of moisture seeping downslope from the upper soils. These soils remain moist longer and dry out less frequently and less completely; this leads to an increasing degree of hydration of iron. The red colour then changes to a brown or yellow one; the hydrated iron oxides are mainly limonite and goethite. The colour changes are not sudden; there is a gradual change from the original reddish-brown of the upper soils to orange-brown and then to yellow-brown and sometimes brownish-yellow on the lower slopes.

On the lowest slopes, where drainage can be very poor and where part or all of the soil profile is waterlogged, reduction of iron and other soil compounds takes place. Under these conditions, bacteria obtain their oxygen from the oxygen-containing compounds and these are then reduced to other compounds. These waterlogged soils are usually bluish-grey, greenish-grey or even neutral grey in colour. In that part of the soil profile where the water-table fluctuates mottling is likely to be produced.

Thus, differences in drainage are responsible for the gradual colour changes that are frequently seen in catenas. These drainage differences can be due to a variety of factors. In parts of Indiana the hydrologic sequence has been shown to be correlated with surface slopes in medium-textured materials, with differences between land and water surfaces in porous materials, with the permeability of fine-textured materials and with flooding patterns in alluvium (Bushnell, 1945). But in each case the gradations within the catena can be related to an oxidation–reduction balance in the same way as described in the West African example.

Three different factors are clearly important in determining these sequences. The surface form of the slope is obviously important but so too is the form of the base of the weathered rock or regolith. The form of this weathering front is largely controlled by the type, intensity and orientation of joints in the bedrock. This weathering front can be extremely variable and the relationship between the weathering front and the slope surface is of basic significance. To these two factors must be added the form of the water-table. These three factors are connected in

a highly complex way but are all of great pedogenetic and geomorphological significance. Whereas the form of the weathering front changes very
slowly with time and the surface form somewhat more rapidly, the water-
table is subject to seasonal, annual and long-term fluctuations and also to
changes imposed by the gradual change of the other two factors. The
concept of slope thus needs to include a full appreciation of all these factors.

One of the ways of portraying catenas is by a series of soil-profile
diagrams representing the soil at different positions thereby allowing
the changes to be seen clearly (Figure 3.2). A slight variant of this has
been used by Williams (1968) (Figure 3.3). The types of changes possible
have been outlined by Young (1976). In the simplest case a single horizon remains unchanged (Figure 3.4(a)), or it may thicken (b) or thin (c).
It may also become deeper (d) or shallower (e). It sometimes ends com-

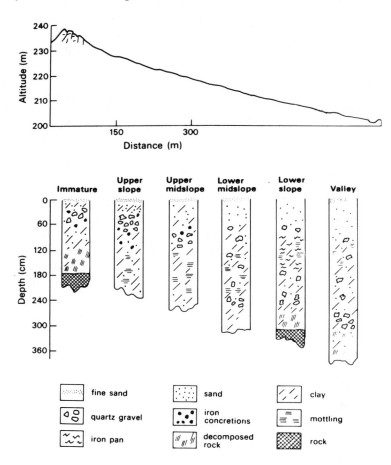

Figure 3.2 Sequential changes in a West African catena (after Nye, 1954).

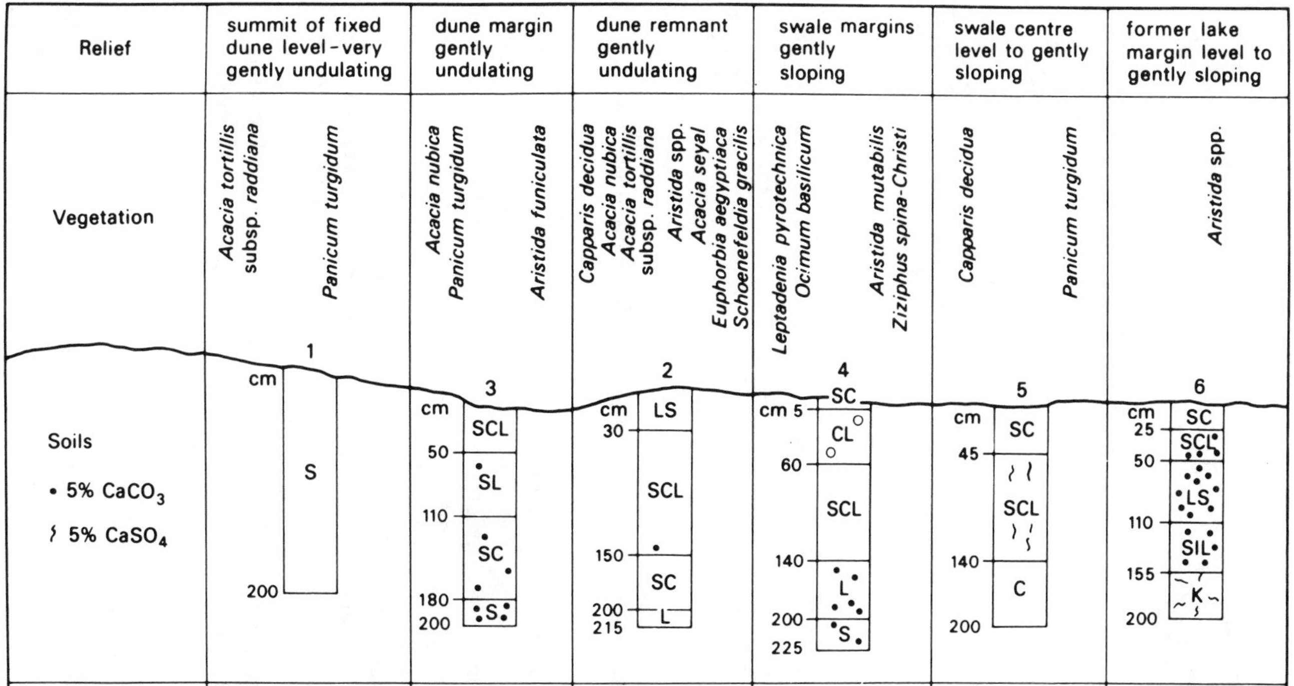

Relief	summit of fixed dune level – very gently undulating	dune margin gently undulating	dune remnant gently undulating	swale margins gently sloping	swale centre level to gently sloping	former lake margin level to gently sloping
Vegetation	Acacia tortilis subsp. raddiana / Panicum turgidum	Acacia nubica / Panicum turgidum / Aristida funiculata	Capparis decidua / Acacia nubica / Acacia tortilis subsp. raddiana / Aristida spp. / Acacia seyal / Euphorbia aegyptiaca / Schoenefeldia gracilis	Leptadenia pyrotechnica / Ocimum basilicum / Aristida mutabilis / Ziziphus spina-Christi	Capparis decidua / Panicum turgidum	Aristida spp.
Soils (• 5% CaCO$_3$, ≷ 5% CaSO$_4$)	1 — S (to 200 cm)	3 — SCL (50), SL (110), SC (180), S (200)	2 — LS (30), SCL (150), SC (200), L (215)	4 — SC (5), CL (60), SCL (140), L (200), S (225)	5 — SC (45), SCL (140), C (200)	6 — SC (25), SCL (50), LS (110), SIL (155), K (200)
Topsoil colour	dark greyish brown	dark grey	v. dk grey-brown	v. dark grey	v. dk grey-brown	dk grey-brown
Subsoil colour	brown	dk grey-brown	dark brown	dk grey-brown	dark brown	v. dark brown
Topsoil % sand	84	77	72	70	57	57
Subsoil % clay	10	19	22	23	32	32
Subsoil ESP	0	32	23	1	53	62
EC in 2nd horizon	0.2	0.9	0.4	0.2	5.8	5.2

Figure 3.3 A dune catena in the Sudan (after Williams, 1968).

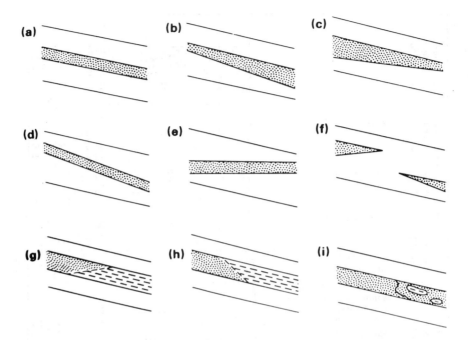

Figure 3.4 Possible horizon changes in a catenary sequence (from Young, 1976).

pletely and a new horizon commences (f) or it may be replaced by another horizon starting from the base (g) or top (h). The final possibility (i) is that a horizon may undergo a gradual change in properties whilst retaining its identity and continuity.

If all the horizons are analyzed in this way three common situations can be identified: first, situations in which there are no downslope changes in successions, depths or properties of horizons; secondly, there may be parts of the catena where one or more horizons undergo gradual change; and thirdly, zones where rapid changes take place leading to substantially modified horizons over short distances. If the changes observed in catenas are analyzed in this manner, soil and slope processes can be related in a more meaningful way.

CATENAS ON SITES OF GEOLOGICAL DIVERSITY

There has been frequent discussion as to whether catenas should be restricted to sequences on one-parent material. In the United States severe limitations are placed on the degree of variation permitted in the parent rock (Watson, 1965). But this is an unrealistic limitation since parent material differences in the catena can occur even though the underlying geology is uniform. The underlying rock is often only the direct parent material of some of the soils, usually the upland ones. The rest of the catena will have developed in transported materials which, although they have been initially derived from the underlying rock, are now composed of weathered and partly-weathered materials which have been transported and possibly sorted. The restriction to similar materials would mean that soils on colluvial deposits would be placed in a separate catena from soils on the adjacent upland. Clearly, this is unrealistic. Hole (1976), overcomes these problems by defining a catena as a group of soils developed from similar initial materials.

But this does not cover the situation where different geological formations outcrop on a single slope. Milne was aware of this problem as the following statement shows:

> Since the first recognition of these catenary associations, it has become apparent that we have to deal with two classes of them. In one, the parent material does not vary, the topography having been modelled out of a single type of rock at both the higher and lower level . . . . In the other kind, the topography has been carved out of two superposed formations, so that the upper one is exposed further down the slope (Milne, 1935b, p.346).

Many examples exist of this second type of catena. In parts of northern Nigeria the landscape consists of sandstone- and ironstone-capped flat-topped hills with steep scarps rising above an extensive undulating sandstone plain. The flat summits are capped with iron-impregnated sandstones or ironstones with little soil. The steep slopes possess a shallow layer of loamy soil over a rubble of sandstone and ironstone. The rest of the area is covered by deep orange-brown to red sandy clays with a brown mottled sandy-clay loam in the depressions.

More complicated situations occur where slopes intersect a variety of rock types such as the slopes developed on dolerites and shales described by Sparrow (1966). Each rock type may then develop its own catena, but since the upper rocks frequently fail to extend far enough downslope to reach poorly-drained sites, the lower catena members are rarely present. If some recurrent pattern of parent material change with surface morphology exists this may be represented as a catena across a relief transect, as in the Chiltern Hills, England (Figure 3.5). The relationship between parent material and soils is complicated here by plateau and valley drift. Thus, the brown calcimorphic soils occurring on the moderate (8–15°) slopes are associated with colluvial or solifluction deposits containing varying proportions of loess-like limon and earlier formed clay-with-flints mixed with frequent chalk (Avery, 1958; Ollier and Thomasson, 1957).

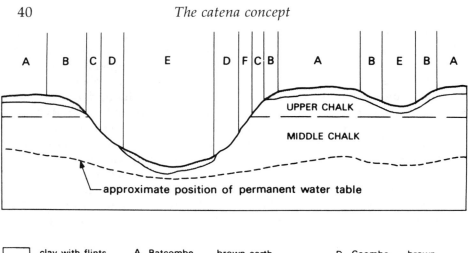

clay with flints
Drift Head

A Batcombe   brown earth
B Winchester brown earth
C Wallop   brown calcimorphic

D Coombe   brown
             calcimorphic
E Charity brown earth
F Icknield rendzina

Figure 3.5 Relationships between soils, topography and geology in the Chiltern Hills, England (after Avery, 1958).

Clearer relationships between parent material and soil types can be found in West Africa (Figure 3.6). Alternating bands of softer schists and harder gneisses and quartzites form a series of ridges and depressions. A thick ironstone cap (A) overlies the quartzites and quartz schists of the ridges. Soils at (B) contain quartz and ironstone fragments and may possess indurated subsoil horizons. Shallow pale-brown gravelly loams (C) overlie weathered schists often with a stone line of quartz or ironstone gravel. Soils (D) on the gentler lower slopes are brown to pale-brown gravelly sandy loams over mottled weathered schist, while the low ridges (E), underlain by gneisses and pegmatites, only possess very shallow soils.

In many instances the complex interaction between topography and geology makes it difficult to predict what type of soil will be found.

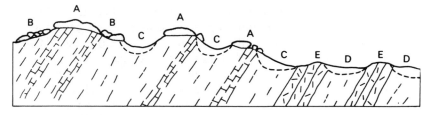

Figure 3.6 West African catena developed on alternating bands of hard and soft rock (after Ahn, 1970; Pullan and De Leeuw, 1964).

Nevertheless, restricting catenas to one-parent material situations is an unjustifiable limitation but it is incorrect to apply the concept to situations where soils are found, *in situ*, corresponding more or less exactly, with the underlying geological pattern, since in this case the effects of 'normal' erosion are absent.

CATENAS AND TIME

The essence of catenas, as discussed above, is the relationship between soils and topography expressed in terms of slope angle and position. A certain amount of time is required before soils become sufficiently differentiated for a catenary sequence to emerge. There is often the implicit assumption that soils have reached some sort of steady-state condition but it is notoriously difficult to decide when this condition exists especially when geomorphic processes are also involved. Curves for the build-up of most soil properties are fairly steep to begin with but after a time they flatten, indicating little change thereafter. But the time necessary to reach steady state will depend on the soil property, parent material and the kind of soil profile developing. 'A'-horizon properties form rapidly whereas 'B'-horizon properties form more slowly. Thus, it has been estimated that podzols in Michigan have taken more than 3000 but less than 8000 years to form (Franzmeier and Whiteside, 1963) while laterites might date from the Tertiary or early Quaternary periods. Much of our knowledge of the time factor in soil formation has been obtained from dateable glacial and volcanic deposits. With time, carbonates are leached out of calcareous till, and iron is oxidized to greater depths with an associated change of pH (Bushnell, 1943; Crocker and Major, 1955).

Therefore, any analysis of catenas should take account of the past physiographic history of the area. As well as the soils changing with time the landforms would also have been changing, and for detailed catenary sequences to exist relationships need to remain reasonably stable. Thus, the catena in any one locality is a complex interaction of landforms, soil and time. Therefore, the catena must be regarded as a dynamic phenomenon with a dimension in time and can be seen as an essential part of the processes of erosion and deposition. With this in mind, some examples of catenas from different climatic environments are now examined.

CATENAS IN DIFFERENT CLIMATES

In their detail all catenas are different but there are sufficient similarities to warrant a preliminary grouping on the basis of world climatic types, notwithstanding the doubt that has been cast on the concept of soil zonality. Thus 'on examination of the catena–climate relationships of the world, we find a contrast between extreme and non-extreme situations. Distinctive soil–slope relationships occur in the extreme situations dominated by frigid or arid conditions . . . . In all the rest of the world, under non-extreme conditions, the processes of slope erosion, slope deposition and pedogenesis are almost inextricably interwoven' (Ollier, 1976, p.166).

Some of these relationships are now examined in a few distinctive catenas.

Tropical savanna catenas

Tropical savanna catenas exhibit a considerable variety of form but it is possible to classify them as catenas with rock outcrops (inselberg and pediment catenas), catenas with a hard laterite (plinthite) cap and catenas without rock outcrops. This classification, based on the work of Ollier (1959) and Moss (1968) is capable of further subdivision (Table 3.1). Although considerable variations exist, each catena is associated with a particular slope form (Figure 3.7).

Table 3.1 Classification of tropical savanna catenas (after Ollier, 1959; Moss, 1968)

(i) Catenas with rock outcrops (inselberg and pediment)
    (1) with extensive pre-weathering
    (2) without extensive pre-weathering

(ii) Catenas with hard laterite
    (1) hard laterite as an upper slope feature
        (a) with massive laterite
        (b) with concretionary or detrital fragments only
    (2) hard laterite as a lower slope feature

(iii) Catenas without rock outcrops
    may be subdivided on the basis of underlying geology

Inselberg and pediment catenas.

This catenary type is characteristic of much of Africa and is commonly developed on granite where hillslope angles seem to be partly controlled by the basal surface of weathering. The majority of slopes leading away from rock residuals are steep pediments varying from 8–10 degrees in angle whilst the lower part of the catena may or may not possess an alluvial member depending on the often complex geomorphological history of the region.

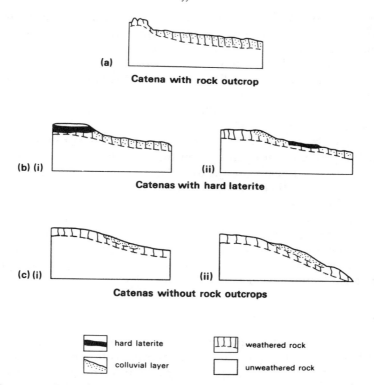

(a) Catena with rock outcrop

(b) (i) **(ii)** Catenas with hard laterite

(c) (i) **(ii)** Catenas without rock outcrops

hard laterite   weathered rock

colluvial layer   unweathered rock

Figure 3.7 Basic features of tropical savanna catenas (after Moss, 1968).

Controversy centres on whether the slope forms are fossil and the result of pedimentation under arid conditions (Birot, 1960), or whether they are being actively developed at the present time (Budel, 1957; Cotton, 1961). Allied to this are arguments concerned with the origin of the deep layers of weathered material and the varying importance of sedentary and colluvial and wash processes (Vine *et al.*, 1954; Charter, 1958; Moss, 1963, 1965). One such catena, in Uganda, has been described by Radwanski and Ollier (1959). This sequence can be divided into three upland and two lowland components, clearly distinguishable at the soil-series level. The upper series (1), called the Buwekula Shallow, occurs as a narrow belt surrounding the rocky inselbergs. Soils are shallow, compared with the other members, in that incompletely weathered rock occurs at shallow depths. The soil is essentially a loamy sand with abundant coarse angular quartz gravel and occasional fragments of feldspar. The Buwekula Red Series (2) occurs on the upper-middle and middle slopes and is usually the most extensive member. As the name suggests, it is mostly reddish in colour and of a sandy clay-loam texture. The Buwekula Brown Series (3) occurs on the low-middle and lower slopes and may be regarded as a variant of (2) in that it has been altered by drainage as influenced by topography. The changes between (1) and (2) are quite sharp whereas those between (2) and (3) are very gradual. The Buwekula Yellow-Brown Series (4) occurs on the slightly raised valley bottoms and valley slopes with the horizons showing evidence of the seasonal fluctuations in the water-table. The Buwekula Grey soil (5) occurs on the valley floor and is often completely submerged in the rainy season.

The interesting aspect of this catena is the evidence it provides not only about the processes in operation on the slope but also about the past geomorphological history of the area. Detailed morphological evidence has indicated that there are three parent materials involved in the catena. The Buwekula Shallow has been derived from fresh or only slightly weathered granite; the Buwekula Red and Brown from intensely pre-weathered granite and the Yellow-Brown and Grey from alluvium derived from the pre-weathered granite. This is interpreted as providing evidence for a 'two-cycle theory' for the evolution of the landscape and pedological features (Ollier, 1959).

Savanna catenas with hard laterite.

Slope form is fairly constant with a flattened upper slope and summit separated from a straight or concave middle portion by a well-marked convexity. The middle slope passes into a fairly straight lower slope of low inclination. The upper portion is associated with thin residual soils; the break of slope coincides with the hard laterite band and the middle slope is dominated by talus derived from the break-up of the laterite. The lower slope is characterized by sedentary soils or by finely-divided talus. This soil and slope sequence is the result of the breakaway retreat of the hard laterite band. Occasionally catenas with hard laterite in a lower slope position are encountered (Table 3.1).

Savanna catenas without rock outcrops.

These are the classic African catenas as described by Milne (1947) in East Africa, Vine (1941) in Nigeria, Charter (1949) in Ghana and Watson (1964–5) in Rhodesia. Nye's (1954–5) study is included in this category because the rock outcrop is only a minor feature. These catenas generally possess smooth convex–concave slopes although stream incision frequently modifies the lower slopes (Figure 3.7). The crest is usually occupied by a dark-red sandy clay with a well-developed structure which gradually changes into a yellowish-red sandy clay with a weaker structure on the steepening convex slope portion. The upper part of the concavity is occupied by a dark-brown sandy clay loam overlying a mottled sandy clay. The middle and lower parts of the concavity, with slopes of $1–2\frac{1}{2}°$, possess a grey sandy-loam or loamy-sand and the valley centres are filled with black hydromorphic clay. Varieties of slope form and soil type within the general framework are produced by different parent rocks. Although a great deal of work has been undertaken in Africa, similar catenas have been described in the seasonal monsoon climates of India (Agarwal *et al.*, 1957; Biswas and Gawande, 1962; Gupta, 1958), in Sri Lanka (Panabokke, 1959) and in Brazil (Askew *et al.*, 1970).

Tropical rainforest catenas

Only a few examples of tropical rainforest catenas have been described (e.g. Delvigne, 1965; Joseph, 1968; Young, 1968). The major distinguishing feature of this catena is the valley and slope form known as sohlenkerbtal, defined by Louis (1964), where strongly convex slopes intersect the level floodplain with little or no concavity. Because of this slope form only a two-member catena normally exists consisting of a freely-drained upper member and a poorly-drained soil on the valley floor. Textural contrast, unlike savanna catenas, is not strong and the valley heads commence abruptly with no distinctive soils.

Catenas in arid and semi-arid regions

Arid and semi-arid conditions form one of the extreme conditions mentioned by Ollier (1976) as being significant in soil formation and therefore catenary development. But these regions also possess other differences which tend to set them apart. In many desert areas erosion is dominant and so soils are thin or non-existent. True soils are found only on the more stable surfaces, such as low-angle alluvial fans (Chapter 7) and pediments, and therefore tend to be distinctive for other reasons. In general the clay content tends to be lower than in humid soils (Haradine and Jenny, 1958) and consequently most desert soils have low exchange capacities (Scott, 1962). Desert soils may also have distinctive mineral suites with montmorillonite and hydrous micas being the characteristic clay minerals (Ismail, 1970). Desert soils are examined in greater detail in Chapter 10.

The relief factor in arid areas is often critical. Cooke and Warren (1973) have noted three factors which set aside dry areas from more humid ones; first, the critical slope angle separating stable from unstable portions of the slope is generally more sharply defined in dry areas; secondly, where the water-table rises above a certain well-defined critical depth it will affect soil properties by capillary rise and saline or alkaline soils will result; thirdly, these two factors often mean that soils on different slopes, and even on different portions of the same slope, may be of different ages. The steeper slopes where erosion is greatest have the youngest soils whilst the lower angled slopes have the oldest soils (Gile, 1967). Repeated phases of erosion and deposition on alluvial surfaces produce more complex situations (Gile and Hawley, 1966).

This means that there is a very close relationship between geomorphic processes, soils and landforms in desert areas. This has been illustrated by Yair (1990) in the northern Negev desert, Israel. Two catenas are examined, one in the Negev Highlands and one on the Hovav Plateau. The hillslope in the Negev Highlands has a high rock/soil ratio, the length of the catenary slope is 68 m and the average slope angle is 18°. The catena can be divided into a rocky section and a colluvial section. In the rocky section outcrops form 60–80% of the surface with shallow lithosols occurring in strips at the base of rock steps. Deeper lithosols are found filling rock fissures and bedding planes of massive limestone. The colluvial section comprises the lower third of the slope and possesses a gravelly soil. The slope on the Hovav Plateau has a low rock/soil ratio, is 95 m long and is also divided into two sections. The rocky section consists of densely jointed, flinty chalk which weathers into cobbles and gravels embedded in a continuous thin-loess veneer. The colluvial section is comparatively stoneless and the transition from the rocky to colluvial section is gradual.

As stressed earlier, one of the main elements of the catena concept is an integration from slope top to base created by the movement of soil and water. In arid situations, such as those described by Yair (1990), overland flow is the major agent in water redistribution. Thus the response to rainfall of the two slope sections is important as is the degree and frequency of flow continuity, passing from the upper rocky slope to the colluvial section. The colluvial soils possess high infiltration rates even after one hour of high-intensity rainfall, but the rocky limestone slopes possess low infiltration rates which drop to zero after ten minutes of rain. Other data substantiate these results (Yair, 1983; Yair and Lavee, 1985). The threshold level of daily rainfall necessary to generate runoff on the rock slopes is 1–3 mm. It is 3–5 mm for stony colluvial soils and 10 mm for stoneless loess soils in the Beer Sheva area (Stibbe, 1974; Morin and Jarosch, 1977). The variability of infiltration rates means that continuous water flow along slopes is a very rare phenomenom, perhaps occurring in only 5% of rainstorms.

The data presented by Yair (1990) clearly show that systematic differences in soil properties can be identified along arid hillslopes. Differences are related to water input–output relationships which are controlled by rock/soil cover ratio. Lithosols develop within rock fissures and trap water leading to salt accumulation and saline soils (Dan and Yaalon, 1982). During heavy rainstorms the liquid limit of the soils can be exceeded (Yair and de Ploey, 1979; Yair and Danin, 1980) and water charged with salts is released. The colluvial sections also exhibit a downslope increase in salinity which is opposite to that observed in semi-arid areas where a downslope increase in leaching intensity is normal.

The movement of water is also critical on the slopes of inselbergs where water flows off the steeper slopes of the rock residuals and accumulates in the coarse alluvium of the plains where it leads to locally deeper weathering eventually seeping out onto gentler surfaces depositing fine sediment (Twidale, 1962). In this way, the weathering products of upper slopes are removed to lower slopes. At the same time there is a change in clay mineralogy from one-to-one clay minerals of the upper leached soils to two-to-one minerals of the alkaline lower soils (Ruxton and Berry, 1961).

For the reasons outlined above it is not surprising that simple catenary sequences such as that illustrated in Figure 3.8(a) occur. In semi-arid regions, with a greater amount of available water, soils tend to be better developed but simple catenary sequences still exist especially where inselberg and pediment landscapes are developed (Figure 3.8(b)). Soils and landscape development are therefore intricately related with new surfaces constantly appearing leading to highly complex situations. Such surfaces have been called pedomorphic surfaces by Dan and Yaalon (1968).

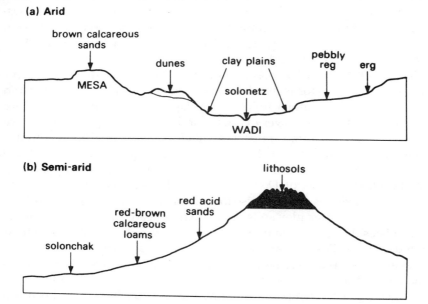

Figure 3.8 Characteristic catenas of arid and semi-arid areas (from Bunting, 1965).

Catenas in temperate regions

It is probably in temperate regions of the world that the application of the catena concept is most in question. There are three factors that make the application of catenas difficult; first, the geological diversity makes simple relationships uncommon; secondly, many temperate areas are covered with numerous superficial deposits; and thirdly, the fluctuations in climate and vegetation systems over the last 10 000 years, and the influence of man on the soils and vegetation have been tremendous and this has upset the evolution of any simple pattern between topography and soil. In order to apply catenas to geomorphic as well as pedologic processes in humid temperate regions, Scheidegger (1986) has stretched the concept to its limits. He discusses fluvial and mass-flow catenas as well as scree catenas beneath rockwalls and argues that the entire landscape is a superposition and juxtaposition of such catena elements. In areas where landscapes have had a complex history, concepts which examine catenas through time are required. The K-cycle system (Chapter 12) of Butler (1959, 1967, 1982) and the scheme proposed by Bos and Sevink (1975), and modified by Valentine *et al.* (1980) attempt to achieve this integration. The most comprehensive approach has been developed by Vreeken (1984) involving soil–landscape chronograms. Such chronograms have time on one axis and erosion, deposition and soil formation across a landscape on the other axis. Chronograms present soil–landform relationships in a space–time diagram and visualize the data within a continuum of processes, concurrent across local landscape systems through time. These points can be illustrated by examples from the British Isles.

Clarke (1954, 1957) has described a mixed catena that occurs on the Jurassic rocks in Oxfordshire. The rock succession is Northampton Sand which forms the hilltops followed by Upper Lias Clay, Middle Lias Marlstone and finally Lower Lias Clay at the base of slopes. At least two soil types may be found on each geological bed providing a complex situation. The problems created by superficial deposits have been mentioned earlier in the case of the Chilterns but many other examples exist. On the Haldon Hills of Devon, soliflucted head deposits mantle the lower gentler slopes and give rise to a distinctive soil series, the Kiddens, comprising of humic gley soils on the loamy head. In other areas of the British Isles the soliflucted deposits form terrace-like features with their own detailed soil–site relationships (Crampton and Taylor, 1967).

Nevertheless, there are situations in temperate regions where the catena concept has been successfully applied. Most of these applications have been in cool temperate regions but catenas have also been described in warm temperate areas (Bricheteau, 1954). The basis of the classification used by the Soil Survey of Scotland is basically a catenary hydrological sequence from top to bottom of the slope producing six separate soils series (Glentworth and Dion, 1949; Glentworth, 1954). A somewhat similar hydrological sequence has been used to describe the soils occurring on the Exeter Shale Hills in Devon (Clayden, 1964, 1971). Well-drained fine loamy brown earths of the Dunsford Series are associated with steep slopes, with more weathered gleyed brown earths and surface-water gleys occupying gentler slopes. Analogous sequences have been described on Exmoor (Curtis, 1971), in the Mendips (Findlay, 1965) and the limestone areas of Derbyshire and Yorkshire (Balme, 1953; Bullock, 1971).

Some of the simplest catena-like relationships in temperate regions occur on glacial tills (Acton, 1965; Brown and Thorp, 1942; Hall and Folland, 1970; Muckenhirn *et al.*, 1949), sand dunes (Matthews, 1971) and loess (Hutcheson *et al.*, 1959; Lotspeich and Smith, 1953). In these circumstances, regular sequences are found with the extent of gleying depending on the topography, position and site drainage. Thus, with care the catena concept can be applied to temperate regions but relationships are often extremely complicated.

Catenas of tundra regions

Tundra regions provide the second of the extreme conditions that Ollier (1976) has suggested produce simple catenary sequences. Soils certainly reflect the climates in which they occur. They are mostly shallow and poorly developed and do not possess well-defined soil profiles. If a traditional A-B-C horizon sequence does develop it may be

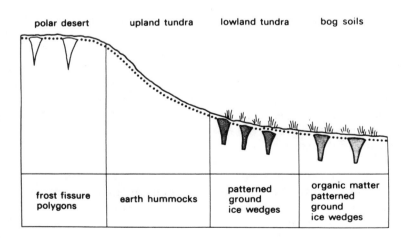

Figure 3.9 Soils and topography of the polar desert zone (modified from French, 1971, 1976; Tedrow, 1974).

subsequently contorted, displaced or obliterated by frost action and solifluction (Tedrow, 1968). Some frost processes are clearly constructive in that they reduce the size of the soil particles and aid the formation of structural aggregates. But many of the processes are destructive as they lead to the physical displacement of the soil. Thus, it is only by grouping frost processes with pedogenic processes that the complexity of polar pedology can be understood. The presence of permafrost also ensures that many of the soils are poorly drained with gleying common.

Catenary sequences that exist simply reflect drainage conditions (Figure 3.9). Polar desert soils are the mature, well-drained soils of the ice-free polar regions (Tedrow, 1966, 1974). They experience neither podzolic nor gley influences; the almost complete absence of vascular plants and the low temperatures and precipitation mean that organic matter rarely enters into the soil system. The distinction between upland tundra and lowland or meadow tundra soils is purely on the basis of drainage. Upland tundra soils are drier and occur on the upper parts of slopes. The characteristic micro-relief consists of earth hummocks and small non-sorted polygons. Lowland tundra possesses more luxuriant vegetation and is often underlain by ice wedges. Bog soils occur in valley bottoms where waterlogged conditions prevent organic matter decomposition and material accumulates to a thickness of two metres. These bog soils are geomorphologically important in that they provide suitable material for palynological analyses and are capable of carbon dating. A fuller treatment of soils and periglacial landforms can be found in Chapter 11.

CONCLUSIONS

It is from the recognition of the systematic relationships between soils, certain landforms and geomorphic processes that the catena concept has been, and still is, of the utmost importance. The real significance of catenas lies in the recognition of the essential processes involved in catenary differentiation. The classification, genesis and geography of soils in catenary association are related to the evolution and elements of the landscape. Thus, the study of catenas has shown that soils cannot be analyzed in isolation from the geomorphological systems of which they are an integral part. Catenas are best developed in tropical and sub-tropical environments, but can, with care, be identified in other areas. Although the concept is easiest to apply in areas with uniform geology, it has been successfully applied to areas of geological complexity. A recent extension of the concept has been the statistical analysis of detailed systematic relationships between slope and position and specific soil properties rather than generalized soil profiles as described in this chapter. This work is examined in the next chapter.

4

Soil relationships within drainage basins

Relationships between soils and slope position and slope gradient, described in the previous chapter, have largely considered general changes to soil profiles. The advent of statistical techniques and the use of high-speed computers enables these relationships to be treated in a more mathematical way. But such procedures require that much greater attention needs to be given to the sampling and measurement schemes utilized to assess these relationships. Soils are extremely variable bodies and the measurement of individual soil properties requires careful consideration. Traditional studies of catenas have tended to examine single slope profiles but, with the aid of the computer, a large number of slope profiles can be handled at one time. This enables the sampling design to consider whole drainage basins and the variability within them of slope form and soils. A synthesis involving slope form, stream system and soils is possible and it is the aim of this chapter to examine some of the relationships that have been discovered.

THEORETICAL CONSIDERATIONS

Many studies have shown that soil properties are related to gradient angle and to length of slope. This is partly the result of the interaction between slope form and the processes of erosion and deposition. The movement of both water and material is governed by geometric configuration of the slope. Thus, these processes can selectively add or deplete the soil of certain physical or chemical characteristics. But relationships between soil properties and slope form are not necessarily simple and predictable. Some constant-angle slopes have been described with greatly varying soil properties and, conversely, soil properties have been found to be reasonably constant over slopes of greatly variable form. Morphology also exerts an influence on the rate of soil formation and the degree of soil maturity. Certain soil properties are related to the stage reached in the evolution of soil profiles. As soils develop, the depth of the profile increases, horizons become differentiated and a characteristic suite of physical and chemical changes can be discerned. But this sequence can be retarded on steep slopes.

A potentially large number of soil properties can be measured and the skill lies in choosing those properties related either to soil maturity or to the formative slope processes. The most useful chemical indicators are pH, organic matter, exchangeable cations and total exchangeable bases, and certain oxides and carbonates. Physical properties such as particle size, moisture content, porosity, plasticity, shear strength, bulk density and aggregation appear to be most important.

It is usually assumed that meaningful relationships between slope form and soil properties will be the inevitable result of any properly conducted study. But realistic correlations will occur only if the processes of soil formation are in some sort of equilibrium with the surface and subsurface processes acting on the slope. No correlations should be expected if the landscape is very young morphologically or if erosive phases are extremely vigorous. A change of climate or a change in the amount and type of the vegetation cover will also upset the equilibrium of the system. Thus a lack of significant correlations may be just as meaningful an indicator of landscape status as the highest statistical relationship.

The choice of soil properties to be measured will be critical. It was noted in earlier chapters how soil properties vary in their rate of change as the soil profile develops. Values of pH and organic content develop and react very quickly to external changes, but clay content and exchangeable bases take a much greater amount of time to achieve equilibrium. Some soil properties will exhibit distinct correlations with slope form whilst others will not, because of this factor.

The other major methodological problem is associated with the measurement of slope form. Slope gradient will vary with the length over which it is measured and this must be borne in mind when considering soil relationships (Gerrard and Robinson, 1971). Slope profiles are essentially curved surfaces and distinct breaks of slope are unusual but schemes, such as those of Ongley (1970) and Young (1971), have been devised to subdivide slopes into curved and straight sections. These methods are also affected by the size of the recording interval used (Gerrard, 1978). The methods also tend to produce differing slope subdivisions when used on the same slope data (Gerrard, 1988a). By far the best methodology would be to produce a continuous record of slope and soil. In this respect it may be possible to produce a continuous trace of soil depth by ground-penetrating radar (Olson and Doolittle, 1985).

Geostatistics offers a technique of examining the spatial variability of soil (Oliver *et al.*, 1989a; Webster and Oliver, 1990). Variograms of soil properties can indicate the scale of spatial dependency. In the Wyre

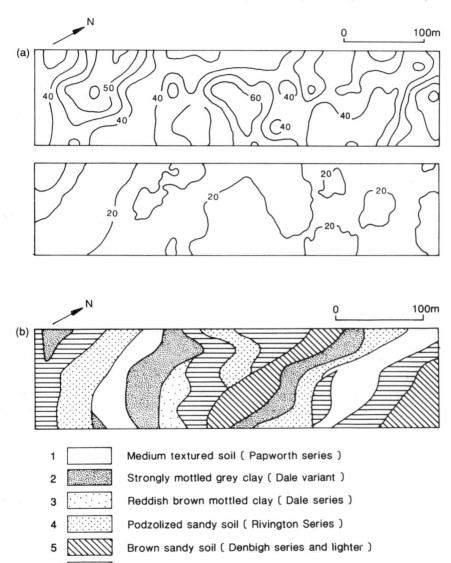

Figure 4.1 (a) Maps of percentage sand and percentage clay in the subsoil made by kriging on a 20 m X 25 m grid; (b) Map of soil types on the two sides of a long transect (from Oliver and Webster, 1987).

Forest, Worcestershire, variograms show semi-variances that increase with increasing lag to about 45 m beyond which they fluctuate about constant values (Oliver and Webster, 1987). Using variograms allows soil properties to be mapped by kriging (Figure 4.1(a)). On Chalk slopes of the South Downs, England, soil depth exhibits a spatial dependence of about 12 m (Gerrard and English, 1990). The variogram assumes that variation is continuous but, if there are marked topographic changes or sudden variations in parent material, transitions are likely to be quite sudden. Such transitions have been noted in the British Isles on the Jurassic scarplands (Webster and De La Cuanalo, 1975), in north-east Scotland (McBratney and Webster, 1981) and in the Thames Valley (Webster, 1973). Boundaries can be located using split moving window and maximum-level variance (Webster, 1973; Hawkins and Merriam, 1973, 1974). Figure 4.1(b) shows one such map for the Wyre Forest produced by these methods. The application of fractals to soil studies also offers a possibility of describing the spatial variation of soils (Burrough, 1981, 1983; Culling, 1986). Techniques such as these may enable soil properties and slope form to be analyzed at relevant scales. The measurement of slope form is necessarily an abstraction and the gradient angles obtained are a function of the techniques employed.

A more intangible problem concerns the distances over which slope gradients should be measured. Soil is usually sampled at one point whereas slope angle is measured over a set distance on either side of the soil-sampling site. The question that needs answering is, over what distance should one expect the soil property to be related to the angle of the slope? The answer to this question hinges on the sensitivity of the system and is reflected in the spatial variability of the soils. Highly sensitive soil systems may react to microscale changes in surface morphology over distances as short as one metre. Less sensitive soil systems will be related to larger landscape components. A lack of correlation between soil properties and slope form may be because the measurement of slope form has been on the wrong scale.

The majority of studies have concentrated on slope profiles as the main sampling units. Soil properties have then been measured systematically along these lines. An alternative scheme is to identify one component of slope form, such as the steepest section or the crestal area, and to sample that section systematically within a drainage basin. Ideally, the two schemes should be integrated to provide a comprehensive account of spatial relationships. To achieve this it is important to distinguish the types of slopes that exist in the landscape because water movement and soil processes are likely to vary with slope type. The most straightforward division is between valley-head slopes, spur-end slopes and valley-side slopes. Valley-head slopes are concave in plan form, spur-end slopes are convex in plan and valley-side slopes are generally straight. Slopes may be further subdivided on the basis of their positions in the drainage basin thus allowing slopes and soils to be related to basin characteristics if necessary.

DRAINAGE BASINS AS PEDOGEOMORPHIC UNITS

The unitary features of form and process exhibited by drainage basins have long been recognized. In 1899, Davis was arguing that, although the river and hillslope waste sheets do not appear to resemble each other, they are only the extreme members of a continuous series. The topographic and hydrologic unity of drainage basins provided the basis for the morphometric system of Horton (1945) which was modified and elaborated by Strahler (1964). Drainage basins provide convenient and usually unambiguous topographic units which can be subdivided on the basis of stream characteristics. This enables a nested hierarchy of both slopes and basins to be established. Drainage basins are also functioning open systems with respect to inputs of precipitation and energy and outputs of water and materials. The most widely used technique for describing the status of streams is Strahler's (1952) modification of the ordering scheme devised by Horton (1945). Fingertip channels are specified as order 1 and where two first-order channels join, a channel segment of order 2 is formed. When two second-orders join, a third-order segment is formed and so on. Although alternative ordering schemes have been devised, such as that of Shreve (1966), the Strahler system is still the commonest means adopted for differentiating streams within drainage basins.

Many relationships have been established between stream order and slope characteristics and, as soils and slopes are closely related, it would be expected that soils and stream order would be related. Strahler (1954), working in the Verdugo Hills, California, established that a general relationship exists between mean and maximum valley-side slope angles and the gradient of the basal stream. Also, undercut slopes were steeper than slopes whose bases were protected by talus and slope wash. A high level of association between stream, slope and soil variables was also reported by Chorley and Kennedy (1971). Again, a major difference was found between slopes that were being actively undercut and the opposing slip-off slopes. Most of the significant correlations on the undercut slopes were within the geometry group of variables implying that one of the most important factors was the rapid transport of debris to the stream. On slip-off slopes, there were numerous strong links between soil and vegetation characteristics and features of the slope profiles. The conclusion reached was that on these slopes a more stable, debris production (weathering) situation existed.

In large or deeply-incised drainage basins aspect-induced differences in microclimate and hillslope processes may occur which will affect soil patterns. Churchill (1982) has demonstrated this in the White River Badlands of South Dakota. North-facing slopes were prone to relatively high-magnitude, low-frequency mass movement failures, whereas south-facing slopes experienced more frequent but smaller movements. South-facing slopes were drier and subject to more frequent and intense episodes of wetting and drying. Higher moisture levels on north-facing slopes lead to more rapid weathering and thicker regolith covers. Clearly this will have an influence on soil patterns. In the Allegheny Plateau of Ohio, soil on southwest-facing slopes have thinner A horizons and more developed B horizons than soils on the northeast-facing slopes (Finney et al., 1962), and in the southern Appalachian Mountains, kaolinite is dominant in soils of the north-facing landscapes whereas gibbsite is the major clay mineral in soils on south-facing landscape elements (Losche et al., 1970). Soil differences related to aspect have also been noted in the Cumberland Plateau of Tennessee (Franzmeier et al., 1969). A comprehensive review of aspect-related differences in soils has been provided by Birkeland (1974, 1984).

The three-fold interaction within drainage basins, of stream–slope–soil, is a complicated combination of process–response systems. One of the simplest methods of unravelling this complexity is to examine slope–soil relationships and then to place these slopes, and their associated pedological and geomorphological processes, in a drainage basin framework.

RELATIONSHIPS BETWEEN SLOPE ANGLE AND SOIL PROPERTIES

Considerable variety of slope form makes comparison between soils and slopes extremely difficult. For this reason, the majority of studies have been undertaken on simple slopes comprising an upper convexity, a maximum segment and a lower concavity. One of the earliest studies was by Norton and Smith (1930) on loess soils in Illinois, where they discovered an inverse relationship between slope angle and depth of the textural B horizon, and correlations between slope and soil structure, texture and consistency. Young (1963) has established several correlations in the British Isles; regolith thickness remained constant or increased slightly across convexities and usually increased downslope on maximum segments. But, Young found no systematic change in the degree of particle reduction over either the convexity or maximum segment, except that the proportion of large stones increased on slopes over 25°. Young also found that regolith thickness and degree of particle reduction increased across concavities. This has not always been found in other studies, but it is clear that there is a fundamental difference between soil relationships on convex and straight units and on concavities.

This distinction has been stressed in a series of papers by Furley (Furley, 1968, 1971; Whitfield and Furley, 1971). High correlations were found between soils and slope angles on convex elements and maximum segments but relationships on concave elements were much poorer. Generalized relationships for the convex and maximum slope zones are shown in Figure 4.2. Acidic soils showed a decrease in pH with increasing gradient angle whereas the relationship was reversed for calcareous soils. All soils, both acidic and calcareous, showed an inverse relationship between gradient and carbon and nitrogen content. For calcareous soils, percentage silt and clay declined with increasing slope angle.

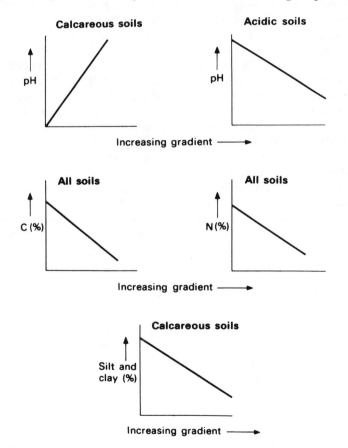

Figure 4.2 Generalized relationships between soil properties and slope angle (from Furley, 1968).

These ideas have been examined on slopes in the Cotswolds, (Jordan, 1974; Gerrard, 1990b). Most of the slopes have simple convex–concave form and are ideal on which to test soil relationships. Soil properties analyzed were pH, moisture factor, organic carbon, total nitrogen, cation exchange capacity, total exchangeable bases and various particle-size grades. The results of regression and correlation analyses showed that only total and upper slopes possessed any significant correlations. Not only were relations more significant on upper slope portions but, in some cases, the form of the relationship was completely reversed on lower portions. Thus, moisture factor decreased with an increase in gradient on upper slopes but increased with increasing gradient on lower slopes. Also, the impression gained from this study was that gradient was not having a marked effect on soil particle size. This has been found in many previous studies and Furley (1971) suggests that under such circumstances, there is little erosion of the mineral soil occurring and no signs of selective removal or deposition.

These results indicate that changes in gradient over slopes do appear to have a significant effect on the values of certain soil properties, but that this effect is not equally distributed over the entire slope. Most change occurs on upper slopes, especially at or around the convexity. A more constant distribution is found on maximum segments and this may mean that this linear zone is one of transit rather than of erosion. Small changes in gradient have some effect on soil properties and significant associations on upper crestal sections lend support to the theory that soil maturity is most advanced on the gentlest of slopes.

The contrast between convexities and concavities is also manifest in soil depth values on chalk slopes in the South Downs, (Table 4.1). Concave sections show a strong tendency towards a negative relationship, although only two of the correlation coefficients are statistically significant, whilst correlations on convex sections are reversed. Even so, no fixed pattern emerges between soil depth and gradient angle on all the slopes examined.

Table 4.1 Soil depth correlations with slope angle and position on eight chalk slopes (from English, 1977)

Slope	Correlation coefficients concave portion	convex portion	total slope
1	− 0.34	+ 0.27	− 0.31
2	− 0.56	+ 0.37	+ 0.11
3	− 0.53	+ 0.34	+ 0.03
4	− 0.29	+ 0.26	+ 0.03
5	+ 0.16	− 0.09	+ 0.14
6	− 0.69	+ 0.59	+ 0.29
7	− 0.66	− 0.04	− 0.06
8	− 0.74	− 0.13	− 0.30

EFFECT OF SLOPE POSITION

Many of the changes just discussed are also likely to be the result of the relative position of the soil sampling site (Aandahl, 1948). Slope gradient on simple convex–concave slopes is highly correlated with distance, especially if the slopes are separated into convex and concave units. Thus, one would expect soil properties to be related to position. Furley's (1971) results demonstrated that this is especially so if the entire slope is analyzed rather than its separate components (Table 4.2). Of the 48 slopes analyzed, on only seven was gradient angle the dominant factor in explaining soil property variation over the complete slope profile. Slope position was the dominant factor on 34 of the slopes. When slopes were subdivided into upper and lower portions and treated separately, interesting changes occurred. For upper slope portions, slope position and slope angle were equally dominant, but on lower slope portions, slope position was the major factor. Relationships were stronger between slope position and soil properties on lower slopes.

Table 4.2 Dominant factor explaining the variation in soil properties on chalk slopes (from Furley, 1971)

Dominant factor	Total slope	Upper slope	Lower slope
slope angle	7	21	10
slope position	34	21	20
angle and position equal	7	6	17

On the basis of these and other results Furley (1971) has produced a model to depict the distribution of soil properties on slopes (Figure 4.3). The basis of the model is that soluble minerals and exchangeable ions are leached from upper slopes, moved downslope and are deposited at the slope foot. The greatest concentration of certain soil properties, such as organic matter, would be expected on the gentle slopes at the top and bottom of the slope. The different trends on upper and lower slopes seem to reflect different processes: upper slopes appear to be associated with processes of erosion and transport and lower slopes with deposition and transport. This is similar to the non-cumulative and cumulative slopes of Nikiforoff (1949) and the aggrading and degrading surfaces of Crocker (1952). It is also not too dissimilar to the major division in the slope and soils model of Dalrymple *et al.* (1968). Lower slopes exhibit greater variability which may reflect erratic zones of deposition.

The junction between the two slope zones is not always sharp but is often diffuse and can oscillate in an irregular manner across the slope according to local variations in microrelief. It is also likely that the junction will fluctuate and migrate up and down the slope in time, with fluctuations in the balance between erosion and deposition. The com-

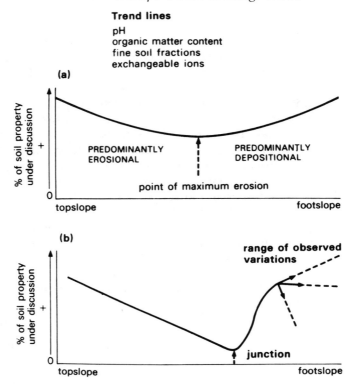

Figure 4.3 (a) Theoretical distribution of soil properties with slope position and (b) the distribution of certain soil properties on chalk slopes (from Furley, 1971).

mon assumption made is that concave slope portions are accumulation zones and that relationships, as discussed above, are the result of deposition. However, this implicit assumption has been challenged by Young (1969). An accumulation zone is, by definition, an area where a net gain of material occurs, but most concavities, where examined in detail, are found to be bedrock features and not accumulations over sharp rock junctions. This means that soil relationships must be the result of differential deposition and removal on a transit slope rather than gradual deposition.

In a later study, Anderson and Furley (1975) examined the relationships between surface properties of chalk soils and slope form using principal components analysis. Five line transects were sited over grass-covered, convex–concave slopes in the Berkshire and Wiltshire chalk downs. Twelve soil properties were used. The first two components in the analysis accounted for approximately 70% of the total variance in the soil properties. The first component comprises an organic matter factor and a soluble constituents factor. The second component seems to reflect particle-size variation. The distribution of the component scores indicates that properties associated with the organic matter factor diminish fairly evenly downslope and those associated with the soluble constituents increase downslope. The distribution of scores for the second component reveals a more abrupt increase in the finer soil fractions at sites immediately downslope of the maximum gradient. The concept of the junction, as expressed earlier, was that it occurred where dominant erosion was superceded by dominant deposition. This analysis suggests that, apart from the fine soil fraction, most soil properties alter gradually over the slope related to pedologic and geomorphic processes which act in a diffuse manner.

Somewhat similar patterns were found on steepland soils in the Wanganui district of New Zealand by Campbell (1973). Four distinct soil units were determined related to different slope elements, namely ridge, intermediate steep slope, eroded slope and accumulation slope. The B horizons of the ridge soils showed greater morphological development and greatest increases in clay while the eroded slope soils were the least leached and weathered. Profiles of accumulation slope soils showed the effects of mixing of parent materials through downslope movement and accumulation and some of their properties showed the effects of seepage of moisture from higher slopes. A similar type of variation, but without an eroded slope unit, was identified by Ives and Cutler (1972).

In geomorphologically more complex situations simple slope sequences of soils are interrupted by rejuvenation and valley downcutting. Successive slope sequences of soils, superimposed on pre-existing surfaces, will be formed. Steepland soil variation patterns will be closely linked with the development of geomorphic patterns and a considerable degree of steepland soil variation will occur. The view that soils on steep slopes are continuously being rejuvenated by creep and colluvial movement downslope has been challenged by Campbell (1975). He has shown that steepland soils form in relatively stable regimes and soil rejuvenation is periodic rather than continuous. This, again, points to the existence of thresholds in soil–landform relationships.

The only situation where lower-slope portions will be pure accumulation zones is in areas of interior drainage or small closed depressions. Examples of such depressions are described later on glacial deposits. Material moves off the slopes but is not evacuated and builds up on lower slopes. Footslopes and toeslope sections, therefore, encroach on backslope sections and concave elements come to dominate the slope system. One such example has been described by Walker and Ruhe (1968). Soil distribution patterns are concentric in plan about the centre of the depression. Sediment and soil trends along radial slope profiles plot as sinuous curves with the degree of sinuosity relating to slope form. This produces the type of relationships shown in Table 4.3. The nature of the sorting mechanism is shown in the change of F : C ratio, which is the ratio of the percentage in the 16 to 250 µm range to the percentage in the 250 to 2000 µm range for the whole of the superficial layer. These soil–slope relationships were best expressed as polynomials. A review of soil–slope functions by Yaalon (1975) has shown that many other polynomial functions have been established. In some instances polynomials provide a better statistical fit, but not always. However, they are often more difficult to interpret in terms of the processes acting on the slope. Whichever method is used, relationships between soil properties and slope form and position are impressive.

Table 4.3 Soil property trends in relation to slope profile components in a closed system on calcareous drift (from Walker & Ruhe, 1968)

| Property | Slope-profile component | | | | |
	summit	shoulder	backslope	footslope	toeslope
mean particle size	minimum	maximum	minimum to maximum	decrease	minimum
F : C ratio	maximum	minimum	maximum to minimum	minimum	increase
organic carbon	maximum	minimum	increase	increase	maximum
depth to carbonate	maximum	minimum	maximum	decrease to zero	zero

SOIL–SLOPE RELATIONSHIPS WITHIN DRAINAGE BASINS

In the general discussion earlier in this chapter two strategies for examining spatial relationships of soils and slopes were suggested. One strategy was based on the slope profile as the unit of study and the previous section has outlined some of this work. But, this strategy is very time-consuming and few studies have involved more than about 20 slope profiles. The alternative strategy is to concentrate on specific slope components and to sample these, and their associated soils, throughout the drainage basin.

This was the strategy employed by Gerrard (1990c) on Dartmoor, southwest England. Dartmoor is a granitic, upland plateau with an average elevation of about 400 m but rising to over 650 m in places. Dartmoor has never been glaciated but was subjected to intense periglacial activity (Gerrard, 1988b). Surface instability is rare and dominant processes are throughflow, both concentrated and diffuse (Ternan and Williams, 1979; Willams *et al.*, 1984). Soil–landform relationships were examined at four spatial scales:

1. General relationships across the entire landscape;
2. Relationships within a single drainage basin;
3. Soil–hillslope relationships;
4. Soil relationships on individual landform units.

At the general landscape level reasonably specific soil–site relationships occur. Upper-valley slopes are dominated by iron pan stagnopodzols (Histic Placaquept) and ferric stagnopodzols (Typic Placaquod). Cambic stagnohumic gley soils (Histic Cryaquept) dominate the main plateau surfaces and humic gley soils (Histic Cryaquept) are found in basin, flush and valley bottomland sites. The lower moorland slopes are occupied by humic brown podzolic soils (Typic Fragiorthod).

Relationships within a single drainage basin were tested in a fourth-order basin (Gerrard, 1982). Considerable variability in soil occurs. Fourth-order slopes are dominated by iron pan podzols but substantial variation in profile form exists over very short distances. Greatest variability is exhibited by the depth of the A–B horizon junction and there seems little relation with slope form or position. Soil on third-order slopes is even more variable. Soil profiles are stonier and humic horizons are thinner than those on fourth-order slopes. Soil on second-order slopes is shallower and stonier and often possesses composite sequences in which thin organic layers are sandwiched between deposits of gravel. Thus, some broad generalizations can be made at the drainage-basin level but the soil profiles also indicate that relationships between soil, gradient angle, slope position and geomorphic processes are complicated.

Soil–slope relationships were examined on six long valley-side slopes. No coherent pattern emerged for relationships between soil properties and gradient angle (Table 4.4). Each slope appears to be operating as a separate system and few valid generalizations can be made. Relationships between soil and slope position are as equally confusing. On some slopes there are marked changes in some soil properties but no changes in surface gradient. One possible explanation is that there are marked changes in conditions below the soil surface, such as depth and nature of the granite weathering front (Gerrard, 1989a). Lack of systematic variation of soil along slopes suggests that slopes have not been acting as integrated systems. Many appear to be formed of small segments linked by irregular or curved sections in a periodic manner (Oliver *et al.*, 1989b). Analysis of soil on some of these segments shows that relationships involving gradient angle can be established.

Thus, there is some justification for treating the drainage basin as a pedogeomorphic unit but there are few significant relationships at the individual slope level. Order exists at the individual slope-segment level but not on entire slopes. Consistent relationships will only occur if soil has sufficient time to achieve steady-state conditions. Dartmoor slopes have suffered considerable disturbance over the last 10 000 years (Gerrard 1989b, 1991c) and steady-state relationships have not been achieved. This Dartmoor study has shown that slope morphology is a poor indicator of environmental changes and is a good example of pedology indicating geomorphological history.

Table 4.4 Summary of relationships between soil properties and gradient angle and slope position on Dartmoor, England (from Gerrard, 1988c)

Soil property	Relationship			
	Gradient angle		Slope position	
pH	2 positive	4 negative	4 positive	2 negative
Loss on ignition	2 positive	4 negative	4 positive	2 negative
Moisture content	3 positive	3 negative	5 positive	1 negative
Percentage gravel	3 positive	3 negative	2 positive	4 negative
Percentage sand	5 positive	1 negative	3 positive	3 negative
Percentage silt-clay	2 positive	4 negative	5 positive	1 negative

In New Zealand, soil–landform relationships have been studied within small drainage basins to examine contrasts in soil development, soil distribution and erosion history along an east-west rainfall and elevation gradient across the Southern Alps (Tonkin *et al.*, 1981; Tonkin, 1985; Basher *et al.*, 1985, 1988). In the eastern front-range region, Doctors Range and Coopers Creek were ice-free during the Late Pleistocene, and possess V-shaped valleys with long convexo-rectilinear slopes and rounded summits. The eastern basin-and-range subregion was represented by Dry Acheron stream and a tributary of Ryton River. The upper Ryton valley contains a series of moraines. Valleys vary from V- to U-shaped, with convexo-rectilinear slopes and narrow rounded summits. The Cropp valley in the western region was occupied by ice in the early Holocene and glacial landforms have since been modified by erosion.

Soil patterns in both Doctors Range and Coopers Creek are similar and show good relationships with basic slope units (Table 4.5). Coopers Creek was deforested one hundred years ago and palynological evidence indicates that Doctors Range may have been forested as recently as 2620 ± 49 years BP (Moar, 1970). Profiles are simple and buried soils are rare. In Doctors Range, shallow A/R soils dominate nose slopes and interfluves, whereas deeper A/Bw/R and A/Bw/C soils dominate at Coopers Creek. A/Bw/C soils occur in hollows, on all aspects in both areas. Three soil catenas were identified, one for weathering-limited nose slopes and two for transport-limited nose and hollow slopes (Table 4.6). Both Dry Acheron stream and the Ryton River area were formerly forested. Most slopes are mantled with colluvium and loess. The coarse-slope deposits are thought to be Late Pleistocene to mid-Holocene in age with the loess being of Holocene age. Soil–landform relationships in the Dry Acheron Stream differ between the upper and lower valleys. On the southeast-facing lee slopes in the upper valley, soils are developed in a thicker and mixed loess and colluvial mantle and possess either A/AB/Bw/BC/C profiles or eroded derivatives. On northwest-facing slopes there is a vertically-striped pattern of soils developed in scree with A/AC/C profiles. Soil patterns in the more deeply-incised lower valley are more complex reflecting changes following forest clearance by fire. The pattern is a mosaic of A/C, A/AC/C, A/Bw/C and composite and compound profiles.

Table 4.5 Distribution of dominant (> 50%) and subdominant (20–50%) soil-profile forms at Doctors Range and Coopers Creek, eastern front-range subregion Southern Alps, New Zealand (from Tonkin and Basher, 1990)

(a) Doctors Range

Slope units	Dominant soil-profile form	Subdominant soil-profile form
Interfluve and nose	A/R (5–30 cm) (rock outcrops)	
Hollow	A/Bw/R or C (50 – 100 + cm)	A/Bw/R (30–50 cm)

(b) Coopers Creek

Slope units	Dominant soil-profile form	Subdominant soil-profile form
Interfluve and nose	A/Bw/R (30–50 cm) A/Bw/C or R (50–100 cm)	A/R (5–30 cm)
Hollow	A/Bw/C (50 – 100 + cm)	A/Bg/Cg (20–60 cm)

Table 4.6 Soil catenas, Doctors Range and Coopers Creek, eastern front-range subregion, Southern Alps, New Zealand (from Tonkin and Basher, 1990)

(a) Weathering-limited slopes

Summit–shoulder	Backslope	Footslope–toeslope
A/R	A/R	A/R

(b) Transport-limited slopes

Summit–shoulder	Backslope	Footslope–toeslope
A/Bw/R A/Bw/C	A/Bw/C A/Bw/C	A/Bw/C A/Bg/Cg

The tributary valley of the Ryton River can be divided into an upper half, where the valley sideslopes are predominantly unvegetated screes, bedrock outcrops and moraines with patches of alpine grassland and shrubland, and a lower half which was formerly forested and is now a mosaic of grasslands, shrublands and unvegetated areas (Tonkin and Basher, 1990). Soils on moraines are formed in loess and have shallow A/Bw/C profiles. Soil patterns on sideslopes are complex with a mosaic of simple, composite, compound and eroded soils. This complexity of soil pattern, in both Dry Acheron Stream and Ryton River, indicates that the eastern mountain basins are experiencing a partial evacuation of soil and debris mantle regolith storages. The Cropp River has an upper valley of alpine barrens, descending to alpine grasslands and subalpine shrubland and forest. The lower valley is steep and narrow and is in the zone of montane forests. The soil pattern is dominated by A/C or R profiles with A/Bw/C or R profiles under grassland vegetation. The better developed soils occur on stable interfluves and noseslopes at higher elevations.

The Dartmoor and New Zealand studies emphasize that soil patterns in drainage basins reflect the interaction between slope materials, geomorphological processes, slope morphology and position. This has also been demonstrated convincingly by Arnett (1971) in a sixth-order drainage basin in Queensland, Australia. Arnett found that mean slope length, mean angle and mean maximum slope angle all tended to increase with stream order. These are trends which have been found in many areas and reflect the opening out of drainage basins by streams enlarging their catchments. Other morphological characteristics, such as rate of convexity and concavity, were examined. The mean rate of convexity increased with stream order up to the fourth-order and then declined. The relationship between rate of concavity and stream order was more complicated and exhibited marked fluctuations.

Geomorphological processes acting on slopes are largely governed by factors of slope angle, length and curvature and, therefore, they should also be related to position in drainage basins, as measured by stream order. This was found to be so by Arnett (1971). The occurrence of sites on which soil creep was dominant was complex (Figure 4.4). Crestal sites exhibited a rapid increase up to order four where creep was dominant on 60% of the profiles. Creep was generally less important on both maximum slope facets and toeslope sites and decreased in importance with increasing stream order. Slopewash was found to be important on all slope components of lower order slopes but decreased rapidly in importance as stream order increased.

Rill action became increasingly dominant as the order of slopes increased. Also it was always more important on maximum slopes and toeslopes than on crest slopes. No rill action was recorded on crestal

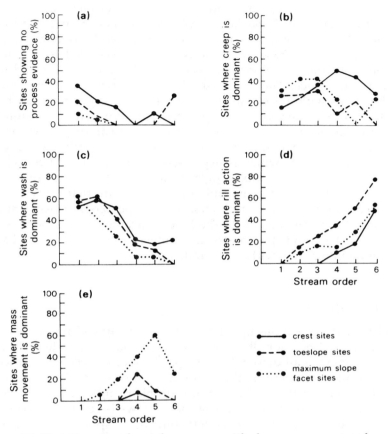

Figure 4.4 Variations in subaerial processes with slope component and stream order, Queensland, Australia (from Arnett, 1971).

sites of fourth-order slopes. Mass movement processes were only important on higher order slopes and then, as would be expected, they occurred mostly on maximum slope areas.

Straightforward relationships also existed between morphological components and certain soil properties. Soil depth, B horizon thickness and percentage clay in the A horizon were found to be inversely related to both maximum slope angle and the rate of convexity. This also means that soil properties will vary systematically within the drainage basin. This was the fundamental point stressed by Conacher and Dalrymple (1977). The relative importance of the components in the nine-unit land-surface model and the sequence in which they occur on a slope will depend on the position of that slope within the drainage basin. As the processes and the nature of the soil cover are all related to position and slope form, soil will also vary within the drainage basin (Table 4.7). This study, by Arnett and Conacher (1973), showed that drainage basin development can be examined in terms of the relations between individual landsurface units. These interactions through time lead to a rational sequence of valley development as drainage basin expansion and integration takes place. This results in the characteristic landsurface unit combinations within each valley or stream order as portrayed in Table 4.7. Therefore, slope form, slope processes and soils create meaningful patterns. This is ample justification for treating the drainage basin as one of the most fundamental geomorphic units and for arguing for order in the landscape; order which is expressed as systematic and repeatable relationships between slopes, soils, position and intensity of stream activity. Conacher and Dalrymple (1977) conclude that the nine-unit model can contribute to both geomorphic and pedologic research at one or more of four distinct levels; the landsurface pedon and polypedon, the landsurface unit, the landsurface catena and the drainage basin. Maximum integration within the landscape occurs at the drainage basin level.

CONCLUSIONS

Good statistical relationships exist between many soil properties and slope form and position. It has not been possible to provide a comprehensive review of all these relationships. All that has been attempted is an assessment of some of the more straightforward of these and to highlight some of the methodological problems inherent in studies of this nature. Relationships are not sufficiently well-established to enable the construction of efficient models, but, there appears to be a fundamental difference between upper convex and lower concave portions. Whether this is simply a distinction between erosional and depositional zones is open to doubt and is something that warrants closer inspection.

Many studies have demonstrated little integration from top to bottom of slopes. Different soil–slope relationships occur on different parts of slopes (Gerrard and Baker, 1990). A number of examples discussed in this chapter cast doubt on the application of the catena concept to slopes in temperate environments. Soils can be shown to be related to landforms but as separate, not integrated systems. The junction between upper and lower slopes where soil–slope relationships are deemed to change has been noted. However, many studies have shown that several such junctions can exist on individual slopes with soil–slope relationships altering at each junction.

The systematic variation within drainage basins, rather than on individual slope profiles, deserves more consideration. Quite often conclusions are based on a sample of relatively few slopes. Any study that tries

Table 4.7 Variations of soil properties within a drainage basin according to the components in the nine-unit landsurface model (from Arnett and Conacher 1973, reproduced by permission of the Geographical Society of New South Wales)

Slope parameters						Soil types			Landsurface units
stream order	length (m)	maximum angle (deg)	mean angle (deg)	convexity (deg/100 m)	concavity (deg/100 m)	crest	centre	toe	unit combinations
1	132	16.7	9.4	16.4	37.0	deep red loam	deep red loam	deep red loam	1 : 5 : 2
2	254	24.1	16.9	30.0	41.0	red loam	red loam	acid red loam	1 : 5 : 3 : 2
3	273	27.4	19.0	37.0	33.0	Shallow loam	skeletal loam	deep red podzol	1 : 5 : 3 : 5 : 8 : 9
4	351	31.4	20.0	45.0	24.0	Skeletal loam	skeletal	deep red podzol	1 : 5 : 3 : 4 : 5 : 6 : 8 : 9
5	396	33.3	21.2	42.0	44.0	skeletal loam	skeletal	alluvial	1 : 5 : 3 : 4 : 5 : 6 : 7 : 8 : 9
6	476	30.9	16.0	40.0	51.0	skeletal loam	skeletal	alluvial	1 : 5 : 3 : 2 : 4 : 5 : 6 : 7 : 8 : 9

to encompass whole drainage basins is valuable and this is one of the reasons why the nine-unit landsurface model has been so influential. The model also possesses the flexibility that enables it to be applied to complex as well as simple slopes.

Although drainage basins can be regarded as fundamental geomorphic units in the landscape, many distinctive landform assemblages occur in other contexts. These, such as erosion surfaces and coastal plains, possess distinctive soil associations which are examined in succeeding chapters.

5

Soils on erosion surfaces

Erosion or planation surfaces have been created by surface or near-surface wear on rock masses. They are reasonably smooth and approximately horizontal planes that cut across structural and lithological boundaries. They can be of any size but the term is usually restricted to large surfaces of low relative relief that are believed to be the end-products of cycles of erosion. The fact that they can be produced by a variety of geomorphological processes, such as subaerial erosion, marine erosion or chemical weathering, has given rise to much confusion and argument. The geomorphological literature is replete with arguments concerning the identification and origin of such surfaces. This is not the place to enter this discussion but the series of papers collated by Adams (1975) provides a fascinating insight into the development of ideas concerning erosion surfaces.

Much of the confusion arises because there is no unequivocal definition of an erosion surface and because similar surfaces can apparently be created by contrasting suites of processes. This is a classic example of equifinality whereby the same end-product apparently can be achieved in a number of different ways. The main types of surfaces are peneplains, pediplains, etchplains and surfaces of marine erosion.

Peneplain was used by W.M. Davis as the name for a gently undulating surface of low relief produced at the end of an erosion cycle by processes of subaerial erosion. Towards the end of the cycle when all the slopes are very gentle the agencies of waste removal must be weak everywhere. 'The landscape is slowly tamed. . . and presents only a succession of gently rolling swells alternating with shallow valleys. . .' (Davis, 1899, p. 497). Relief becomes less and less and an almost featureless plain, showing little sympathy with structure and controlled only by a close approach to base level, characterizes the penultimate stage. The ultimate stage would be a plain with little relief, but perceptible inequalities of 30–50 m would probably still remain. These inequalities in relief will be sufficient to influence the fine detail of the variations in soil distribution.

Davis, although inventing the term peneplain, acknowledges that the idea came from the writings of Powell. Powell has written of mountains as being ephemeral topographic forms that ultimately would be reduced to low near-horizontal surfaces and 'the degradation of the last few inches of a broad area of land above the level of the sea would require a longer time than all the thousands of feet which might have been above it' (Powell, 1876, p. 196).

Since its formulation, the peneplain concept has been heavily criticized, especially by workers whose experience has been in arid rather than humid temperate environments. The first major challenge was W. Penck's *Die Morphologische Analyse* in 1924. His conclusions were based on three assumptions. These were: slopes are established by the downcutting of streams and are steeper the greater the rate of downcutting; slopes once established retreat away from the stream parallel to the original declivity; and steeper slopes are denuded more rapidly than gentle slopes. As far as soils are concerned, original remnants of earlier erosion cycles remain unaffected longer in the Penckian model than they do in the Davisian cycle.

At the same time as Penck was introducing his ideas in Europe, workers in the arid south-west of the United States were describing pediments; gently concave erosion surfaces often mantled with superficial materials of variable thickness that occurred at the foot of the steep mountain scarps. It was soon clear that pediments were being formed by lateral planation by streams issuing from canyons, and by rills cutting at the foot of mountain slopes. Outliers and unreduced remnants were being denuded by weathering and transportation of the debris by rills. Pediments develop by the parallel retreat of mountain fronts and the ultimate cyclic landform is the pediplain, consisting of broad coalescing pediments (King, 1953).

The effects of intense chemical weathering and the advance of the weathering front can produce a third class of erosion surface; the etchplain. The term was used by Wayland (1933) when suggesting that some planation surfaces in Uganda may have been created by the etching of a previously formed surface by deep chemical weathering followed by surface wash. Budel (1957) recognized the role of deep chemical weathering and wash and introduced the term double-surfaces of planation (doppelten einebungsflachen) which are developed as the weathering front descends into bedrock, while surface wash removes the upper weathered zones. Adams (1975) argues that etchplains seem to require the presence, or former existence, of a previously developed peneplain or pediplain. Also, the deep contemporaneous weathered zone on a peneplain, if stripped away, might reveal a surface that could be considered an etchplain. The differences between the various erosion surfaces are probably not as great as many workers suggest.

Large, extremely level uplifted erosion surfaces, have also been attributed to the work of the sea. Except in those cases where marine deposits attest to their origin as uplifted coastal plains, it is doubtful if many erosion surfaces are the result of marine action. To produce a marine-cut surface of any appreciable extent requires a steadily rising sea level for a considerable period of time. A rising sea level is necessary because, without it, marine energy is soon dissipated as the marine bench extends and the cliffs will not be attacked, nor the material moved across the surface. What little is known of previous sea levels suggests that this situation is extremely unlikely. Thus, erosion surfaces created by marine agencies are likely to be limited in extent.

Other specialized, but limited, forms of erosion surface exist. Altiplanation or cryoplanation terraces are formed by the retreat of rock slopes by frost action with the material being removed by solifluction (Reger and Pewe, 1976; Priesnitz, 1988). In common with other erosion surfaces, they cut across different rock structures. Little is known about the precise climatic requirements for the development of cryoplanation terraces but optimum formation seems to occur from 300 m to 500 m below the snowline to a little above it (Richter *et al.*, 1963). A fuller treatment of cryoplanation surfaces can be found in Chapter 11. The term panplanation was introduced by Crickmay (1933) to describe the process of lateral planation by rivers which develop a uniform surface of overlapping rock-cut straths as divides are cut through. But it is doubtful if such a surface has ever been positively identified.

AGE AND STATUS OF EROSION SURFACES

Erosion surfaces take a long time to be produced and during that time external conditions might have changed. These conditions include crustal movements and climate changes, and many erosion surfaces are now isolated from the processes that created them. It must also be borne in mind that erosion surfaces are not necessarily synchronous everywhere. The following classification of erosion surfaces defining their current status has been produced by Adams (1975):

1. Active surfaces are still being shaped by their formative processes;
2. Dormant surfaces are those whose active shaping has ceased temporarily, perhaps by climatic change, and are expected to function again in the near geological future;
3. Exotic surfaces have formed under climatic conditions that no longer exist;

4. Defunct surfaces have been removed from the action of erosive processes by uplift or depression;
5. Buried surfaces have been covered by sediments not related to their shaping; and
6. Exhumed or fossil surfaces are buried surfaces that have been exposed by the removal of a non-genetic cover such as later sediments.

Soils may provide part of the information on which to base the status of any erosion surface. Thus, exotic surfaces can be identified by a soil cover that indicates that there have been gross changes of climate. Also, the type and distribution of soils will make it clear when active erosion surfaces are being formed at the expense of dormant or exotic surfaces, as in many parts of Africa. But analyses of this nature require a thorough understanding of the way erosion surfaces develop and how soils respond to this development.

SOIL DEVELOPMENT AND THE FORMATION OF EROSION SURFACES

It is likely that soil development and distribution will be governed by the different evolutionary patterns of the various erosion surfaces. Mulcahy (1961) has outlined, in general terms, soil development in relation to a landscape evolving by the cutting of successive pediments (Figure 5.1). The landscape shows a series of pediments separated by steep scarps. The lower surface A is developing at the expense of surface B and is, therefore, younger. If soil age is equated with the age of the surface, the soil on surface B will be older and better developed than that on surface A. A stepped succession of soils should ensue and this will also be the situation in a region which is composed of a series of uplifted and partially dissected peneplains. Also, on any one pediment

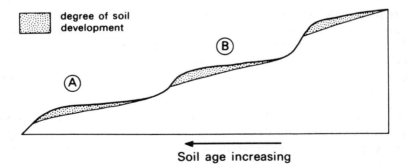

Figure 5.1 Soil development in relation to the retreat of scarps (after Mulcahy, 1961).

the outer edge is older than sites nearer to the retreating scarp. This should mean that soils will be better developed towards the edge of the scarp and less well-developed near the scarp foot, while soils on the scarp face will be perpetually youthful. But a number of factors operate to upset this simple sequential development.

Appreciable soil development will only take place once the surface is no longer affected by severe erosive processes. In the case of pediment formation the whole surface may be affected all the time and not just the areas nearest the retreating scarp. The processes acting on pediments have been described in the following terms by Kirk Bryan (1922). As streams emerge from mountains, large particles are dropped and smaller ones are carried forward. The fine debris washed down by rains is moved forward by rill action to the larger streams, but because the supply of debris is small, the rills are not fully laden and are effective erosive agents, reducing the height of interill areas. The irregularities of the pediments are removed, with the higher parts being eroded and the lower parts infilled with debris.

Differential sorting of sediment on the surface will cause changes in texture, bulk density, permeability and porosity, which will affect the formation of soil. Thus, soil development will depend on the local configuration of the rill network. It also means that soils existing on erosion surfaces will not all be of the same age and may also be considerably younger than the surface itself. The criteria adopted by geomorphologists in the identification of erosion surfaces are not sufficiently critical to detect minor planations, and the ages assigned to major erosion surfaces may bear little relation to the absolute age at any point. But, as Mulcahy (1961) points out, such minor planations and modifications may remove complete soil profiles, exposing weathered or even unweathered material on which soil formation starts afresh. The pattern of soil distribution and age relations is completely altered.

PEDOLOGICAL RELATIONSHIPS

Many of the points discussed above can be amplified with a few specific examples. Some glacial till landscapes of North America show a sequence of stepped levels as a result of multi-cyclic erosion of the landscape. One such area, in Iowa, has been intensively studied by Ruhe (1956). The highest surface is a relict of the Kansas drift plain that has not been changed by erosion since Kansan time. In terms of the classification outlined earlier, this is a defunct surface. But the situation is complicated by the fact that fossil soils or palaeosols (Chapter 12) of Yarmouth–Sangamon age exist on the surface and have been buried by the later deposition of loess. So in places the surface is buried, in other parts the surface has been exhumed by later erosion and the time status of the surface can only be established by a detailed examination of soils and sediments. Morphology alone is not sufficient.

Cut into the upper surface is a pediment of Late Sangamon age (last interglacial) in the sediment of which exists a palaeosol. The low level of the landscape is an Early Wisconsin (last glacial) pediment, cut into the Kansan till, below the Late Sangamon surface. No palaeosol is found on this surface indicating that loess deposition followed very closely after the cutting of the pediment. All these surfaces, and their respective sediments, have been subjected to erosion and sedimentation in Late Wisconsin–Recent time. This has had the effect of exposing the Yarmouth–Sangamon and Late Sangamon palaeosols and surfaces so that these relict soils and surfaces are now part of the modern landscape. Although standard geomorphological techniques allow the general sequence of events to be established, it is the detailed information provided by soils and the pedologist that completes the picture. As surfaces become older total soil and B horizon thickness increase, clay content of the B horizon increases and heavy and light mineral weathering indices increase (Table 5.1). Wrh is the ratio of quartz : feldspar and wrl the ratio of zircon and tourmaline to amphiboles and pyroxenes. The indications are that, although vegetation at the time of soil formation was different on individual surfaces, soil differences are mainly a function of time.

Table 5.1 Soil properties on stepped erosion surfaces in Iowa (from Ruhe, 1956)

Surface	Soil	Thickness of solum (in)	Thickness of B horizon (in)	Clay in B horizon (%)	Soil horizon	Wrh index	Wrl index
Recent	A	15	11	31.2	A	0.79	2.09
	B	32	23	32.2	B	0.92	2.13
	C	29	22	34.6	C	0.68	2.21
	average	25	19	32.3			
Late	D	46	32	50.7	A	1.27	3.06
Sangamon	E	70	56	49.1	B	1.12	2.49
	F	39	29	49.5	C	0.77	2.04
	average	52	39	49.7			
Yarmouth–	G	87	70	51.4	A	2.11	4.85
Sangamon	H	68	44	57.7	B	1.62	3.00
	I	85	62	50.7	C	1.28	2.57
	average	80	59	53.2			

Detailed pedological investigations also enabled Heine (1972) to correct some misconceptions about erosion surfaces in Germany. Although some true Tertiary soils were discovered, many of the supposed soil formations of the Tertiary period, that had been used to explain the morphologic development of the region, were found to be hydrothermal disintegration products. Also, many surface residual deposits and soils showed different age sequences because of the possibility of exhumed surface remnants.

Analysis of soils and weathering products allowed Ollier (1959) to establish the sequence of erosion surface formation in Uganda. Ollier's premise was that present-day soils were formed on pre-weathered rock.

Gondwana surface

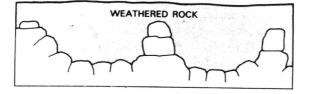

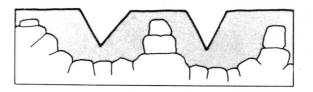

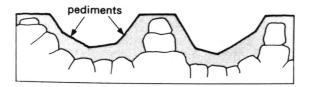

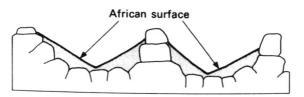

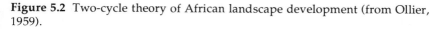

Figure 5.2 Two-cycle theory of African landscape development (from Ollier, 1959).

The sequence of events is outlined in Figure 5.2. Very deep weathering took place below the Gondwana surface to form a thick rotted layer. During the African cycle of erosion much of the weathered rock was removed by the growth of pediments and the parallel retreat of the steep slopes. But, the base-level of erosion during this cycle was reached before all the weathered rock was removed and the surface is largely cut across this rotted rock. The problem was to establish that two cycles had been involved, i.e. a cycle of deep weathering followed by erosion and modern soil formation, and not one continuous cycle of weathering and soil formation.

Evidence used by Ollier (1959) was both pedological and mineralogical. Various features of the soil profile pointed to a two-cycle origin: the presence of stone lines in the soils indicates that a change has taken place in surface conditions; the type of laterite in the profiles also supports a two-cycle theory. If laterite forms in weathered rock, *in situ*, it tends to be massive and vesicular, full of channels which are lined with layers of limonite. If laterite forms in upper, sorted layers of the profile it forms murram pisoliths. The laterites in the exposed profiles appear to be of this second type. Evidence provided by soil profiles certainly suggests a two-cycle origin for the landscape. Mineralogical evidence is less conclusive. Soils formed on upper slopes are rich in mineral species whereas soils on lower slopes are relatively poor in minerals, the indication being that upper soils are formed from the lowest zones of the weathering profile. This could be achieved by greater weathering on midslopes than on upper ones but, taken in combination with the pedological evidence, is strong evidence for a two-cycle theory as proposed by Ollier.

DURICRUSTS AND TOPOGRAPHY

Many of the concepts discussed so far are exemplified in relationships between duricrusts and landforms. Duricrust, as a term, was introduced by Woolnough (1927) to describe chemically indurated material found capping many areas of Australia. Induration is produced by a variety of mineral materials. An all-embracing definition has been produced by Goudie (1973, p.5):

> A product of terrestrial processes within the zone of weathering in which either iron or aluminium sesquioxides (in the case of ferricretes or alcretes) or silica (in the case of silcrete) or calcium carbonate (in the case of calcrete) or other compounds in the case of magnesicrete and the like have dominantly accumulated in and/or replaced a pre-existing soil, rock or weathered material, to give a substance which may ultimately develop into an indurated mass.

Specific definitions for particular types of duricrust have been provided by McFarlane (1983) for laterites, Summerfield (1983) for silcrete and Watts (1980) for calcrete. The nature and distribution of duricrusts can be obtained from a variety of sources (e.g. McFarlane, 1976; Summerfield, 1978; Netterberg, 1980; Bardossy, 1981; Watson, 1982; Wilson, 1983; Goudie, 1985).

Duricrusts can form very rapidly which helps to preserve sensitive landforms such as dunes and alluvial terraces. But the formation of duricrusts requires the interaction of numerous processes, thus quick formation will only occur when all conditions are favourable. The primary elements which make up the duricrust come from at least four main sources: the weathering of bedrock and sediment, inputs from dust and precipitation, dissolved solids in groundwater and plant residues. These elements then require translocation and concentration either by lateral or vertical movements, both upwards and downwards. Then the materials need to be precipitated, helped by changes in chemical equilibria caused by evaporation, temperature changes, pressure changes in the soil air-water system or by the actions of organisms. All these stages have been outlined in detail by Ollier (1978) in a model of silcrete formation. As already seen, the nature of the laterite was a crucial element in Ollier's two-cycle theory. Level or nearly level surfaces in tropical areas are often capped by laterite, and laterite was thought to be a residual formation requiring a landsurface of low relief to allow the downward percolation of surface waters. Steep slopes allow an ingress of water which removes the more mobile soluble constituents and causes the accumulation of the less mobile weathering products.

This concept agreed well with the traditional Davisian cycle of peneplanation as slopes are reduced to a point where vertical water movement through the soil and regolith is substantially greater than runoff. The laterite was assumed to develop with the landsurface and the end-product would be an erosion surface covered with a thick sheet of residual material. However, later ideas implied that laterite was a precipitate and only began forming after the planation surface had been created and after conditions of a stable, but fluctuating, water-table had been achieved. Thus, MacClaren (1906) and Woolnough (1918) both concluded that a level erosion surface was necessary before laterite could form. This idea had far-reaching consequences as far as denudation chronology and the evolution of erosion surfaces was concerned since a thick sheet was regarded as indicative of an erosion surface which had survived for a considerable period of time.

However, more extensive surveys have shown that laterite occurs on quite steep slopes, with angles in the range 7–10° being frequently noted (e.g. Pallister, 1951; Mulcahy, 1960; Trendall, 1962). Laterite has even been reported on slopes up to 20° (De Swardt, 1964). It has been argued that these slopes are not the original form of the surface but are due to post-incision modifications. These modifications may be due to cambering of the margins of mesas, accelerated erosion of the margins of mesas forming a convex waxing slope, or the creation of detrital laterite as a link between high and low laterite sheets.

Studies by MacFarlane (1971), in Uganda, suggest a compromise. MacFarlane regards laterite as a residual precipitate. Groundwater laterite is believed to accumulate as a chemical residuum during the late stages of a downwasting landsurface. The precipitates form within the range of fluctuation of the groundwater table, which sinks as the landsurface is lowered by erosion. Precipitates accumulate as increasingly thick layers in the lower parts of the soil. When downcutting has ceased, the water-table becomes stabilized, the residuum is hydrated and a massive type of laterite forms which has the appearance of a true precipitate.

The relationship between laterite and erosion surfaces is critical and has played a prominent part in the construction of most African denudation chronologies. MacFarlane (1971, 1973, 1976) has reviewed thoroughly the place of laterites in the development of theories of landscape evolution in Uganda. Wayland (1921) originally recognized a simple peneplain, which he called the Buganda peneplain and which he associated with the African cycle of erosion. Subsequently, he recognized two others, an older surface present on the resistant interfluves of southwest Uganda and a younger surface developed in the central lowlands (Wayland, 1933, 1934). More detailed observations of the Buganda surface have shown that it possesses considerable relative relief and that some laterite sheets are apparently continuous through an altitude range of about 150 m. Thus, there is the problem of explaining why laterite, presumed to be a planation surface development, should occur at a variety of altitudes on one of the best examples of a laterite-capped planation surface.

This prompted MacFarlane (1976) to pose a number of questions. Is laterite formation restricted to surfaces of low relief? If the answer is yes, the Buganda surface is not one but two or even three surfaces. Can laterite form on a surface of high relief enabling a single Buganda surface to exist? Can laterite form synchronously at different levels, possibly covering surfaces at different ages? Can a single near-level laterite surface be differentially lowered, creating the impression of two chronologically separate surfaces? Can laterite form independently on separate interfluves to form an acyclic 'apparent peneplain'? To answer these questions the relationship between laterite development and landsurface development needs clarification and, as MacFarlane (1976) concludes, a closer study of laterite type and landsurface type should make a positive contribution to an understanding of landsurface evolution in the tropics.

SOILS ON STEPPED EROSION SURFACES

Periods of extensive erosion surface development and intermittent uplift have produced sequences of stepped erosion surfaces in many parts of the world. The time sequence of surface development provides either an absolute, or relative, means of evaluating soil development. Soil types might be expected to reflect the age and type of the surface and such a sequence occurs in the Willamette Valley, Oregon (Parsons *et al.*, 1970). Seven major surfaces exist (Figure 5.3). The Eola surface is the oldest stable surface in the region and is probably Middle Pleistocene in age; the topography is typically rounded and surface materials are deeply weathered. The Dolph unit exists as dissected flats, underlain by weathered gravel and bedrock, and includes numerous small pediments and straths; probably it is also Middle Pleistocene in age. Calapooyia surface is extensive and extremely level; differences in elevation usually do not exceed a metre; surface drainage is poorly organized and sluggish and the surface may be of marine or estuarine origin; it is probably of Late Pleistocene age.

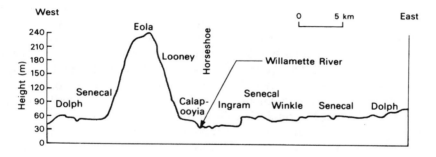

Figure 5.3 Suite of erosion surfaces in the Willamette Valley (from Parsons *et al.*, 1970).

The Senecal surface appears to be a slight modification of the Calapooyia surface occasioned by minor incision of the drainage net; it must also be Late Pleistocene in age. The Winkle surface is one of the most extensive, and is the oldest surface associated with the present drainage systems. Subparallel bar and swale topography of old channels suggests braided overloaded stream channels in some localities and radiocarbon dates indicate an age of approximately 6000 years BP. The Ingram surface includes the higher flood plains of the Willamette River. The lower parts of this surface commonly flood and the oldest radiocarbon date for the surface is 3290 ± 120 years BP. The Horseshoe surface is the lower flood plain of the Willamette River and its major tributaries and is underlain by sandy alluvium and gravel.

Soils developed on the surfaces show interesting sequences. Organic matter distribution with depth shows a general decrease in organic content with increasing age of the surfaces. The weighted mean percentage of organic matter in the A horizon decreases from soils on the Horseshoe to those on the Calapooyia surfaces. It then increases from the Calapooyia to the Eola surface. This may represent greater accumulation of humus under forest whereas soils on the lower surfaces were developed under grass vegetation. Soils of the younger surfaces have high base saturations. Soils with B horizons having clay films are first found on the Winkle and Senecal surfaces and thereafter become more prominent. Variations in soil types are also created by local differences of drainage and moisture status. Organic accumulation appears to be rapid in the Willamette Valley and the A horizons of the youngest soils may contain as much organic matter as soils on the older surfaces. Also, cambic B horizons appear to form within 550 years but the development of argillic horizons requires between 550 and 5250 years.

Differences between soils on stepped erosion surfaces are not always simply the result of age differences, other environmental factors have to be taken into consideration. A series of stepped erosion surfaces occur in the Wahiaua Basin, Oahu, Hawaii (Ruhe, 1975). Although the dominant part of soil associations relates to a specific erosion level, most soil associations cross several levels in the sequence (Table 5.2). Soil association areas are concentrically banded with elevation which also corresponds with climatic zones. Differences in soils are likely to be due to differences in climate and not just to age differences. The same problem of isolating the relevant environmental factors was faced by Young and Stephen (1965) in Malawi. The ferallitic soils of the high plateaux possess a much higher organic matter content in their upper profiles than soils of the lower plains. This sharp increase in organic matter amounts at higher altitudes was explained by the drop in the mean annual temperatures.

Table 5.2 Correspondence of soil associations with geomorphic surfaces, Oahu, Hawaii (from Ruhe, 1975)

Soil	Geomorphic surface (%)							
	5	6	7	8	9	10	11	12
Malokai	30.0	57.6	4.4					
Lahaina		39.8	50.6	9.7				
Wahiawa		2.3	26.8	46.7	18.5	5.8		
Leilehua				6.0	29.6	64.4		
Waipio						85.7	14.3	
Kunia		1.1	21.4	32.4	45.0			
Mahana					15.7	84.3		
Manana						41.7	56.4	1.9
Paaloa							25.0	75.0

These studies demonstrate the problem of isolating the factors involved in soil formation. Erosion surfaces may represent a time sequence but other factors will also have changed through time. Analysis of soils on erosion surfaces and on sloping areas linking such surfaces enables whole soil landscapes, of substantial extent, to be described. One such area where this is possible is in the northeastern United States where soil–landform relationships can be established from the Ridge and Valley Province of the Appalachians across the Blue Ridge and the Piedmont and on to the Coastal Plain. Soils in the Ridge and Valley area have had a very complex genetic pathway, largely the result of climatic change during the Late Pleistocene and Holocene (Ciolkosz *et al.*, 1989). A typical soil landscape of the region is shown in Figure 5.4. The major ridges are usually composed of sandstone and the valleys of limestone or shale. The generally level ridge tops have usually been interpreted as remnants of the Schooley peneplain and lower surfaces as remnants of the Harrisburg peneplain (Thornbury, 1965; Sevon *et al.*, 1983). However, the nature and number of such surfaces has been challenged by recent work (Poag and Sevon, 1989; Sevon, 1989).

Soil–landform relationships have been analyzed by Ciolkosz *et al.* (1990). Soils on the ridge top and shoulder slope areas are deep (> 1 m), well-drained, acid, sandy and have a high stone content. The commonest soils are Dystochrepts (Hazleton) and Haplorthods (Leetonia). Such soils do not show intensive development and are much younger than the age of the parent material. Carter and Ciolkosz (1986) have shown that on the ridge-top upland bedrock is 3 to 4 m deep, while on the shoulder slopes and upper back-slope areas, depth to bedrock is 5 to 7 m. They established that the upper 2–3 m of material is colluvium that has moved downslope and that the lower material is residuum. The upper material has also been cryoturbated during Wisconsin time. Radiocarbon dates of 27 to 28 ka for colluvium in northern West Virginia (Jordan *et al.*, 1987), thermoluminescence dates of 12 to 18 ka from basal loess on colluvium in unglaciated southwestern New York (Snyder, 1988) and a date of 12 ka from basal bog material on top of colluvium in central Pennsylvania (Watts, 1979) adds circumstantial evidence for such an interpretation. Using information on rates of production of soil material (e.g. Alexander, 1985, 1988; Sevon, 1984) and the fact that during Wisconsin times the area was a tundra with permafrost (Ciolkosz *et al.*, 1986b; Delcourt and Delcourt, 1983, 1986), Ciolkosz *et al.* (1990) conclude that the majority of rock weathering occurred before the Wisconsin Period.

The genetic pathway of the ridge top-soils has been complex. A, B and upper C horizons have formed on sandstone colluvium whereas lower C horizons have developed in residual material and are considerably older than the upper horizons. The soil thickness of the ridge top

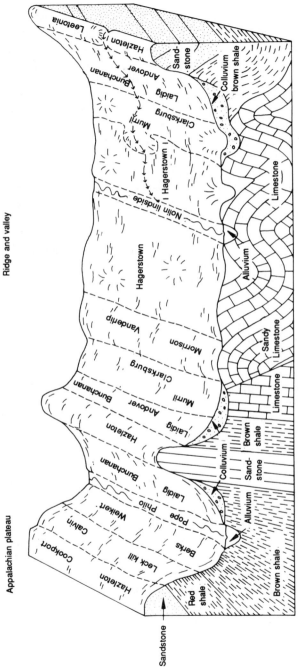

Ridge and valley

Appalachian plateau

Figure 5.4 Generalized diagrammatic soil-landscape relations of the Ridge and Valley and adjacent plateau area in central Pennsylvania (after Ciolkosz *et al.*, 1986a).

Spodosols is about a metre and the soils have taken about 10 000 years to form, according to the criteria of Franzmeier and Whiteside (1963). The Hazleton is the dominant soil on the ridge-tops and the most extensive soil in Pennsylvania (Ciolkosz *et al.*, 1986a). It also occurs on large areas of the unglaciated Appalachian Plateau. The sandy texture and cambic subsurface horizon is largely a function of the sandstone parent material. However, the A and B horizons have been enriched in clay and silt, probably the result of aeolian inputs. It is now recognized that aeolian additions are common in soils (e.g. Smith *et al.*, 1970; Syers *et al.*, 1969).

Soils on the footslopes are deep, acid, well- to poorly-drained, medium to fine textured and most have fragipans. They are developed in colluvium and are classified as Fragiudults (Laidig, Buchanan), Fragiaquults (Andover), Fragiudalfs (Clarksburg) and Hapludults (Murrill). The colluvium has been derived from ridge-top sandstones, sandstone and shale from the backslopes and shale and limestone material from the footslope area. There is a trend of increasing clay and silt content from top to bottom of the footslope. Detailed analysis has indicated an older colluvium, buried beneath more recent material (Hoover and Ciolkosz, 1988). The buried colluvium has bright red colours whereas the upper colluvium is primarily yellowish-brown. The brown colluvium appears to be contemporaneous with the ridge-top colluvium. Soils reflect the texture of the parent material thus those developed in sandstone-shale colluvium are more abundant and tend to be more poorly drained than those developed from limestone colluvium (Table 5.3). Valley soils on limestone are deep, well-drained, acid and clayey. They are classified as Hapludalfs (Hagerstown) and are developed in colluvium. Soils in the shale valleys are moderately deep, well-drained, acid and have silty textures. They are classified as Dystochrepts (Berks, Weikert).

Table 5.3 Relative amount (%) of colluvial soils in various drainage classes in the central Pennsylvania Ridge and Valley (from Ciolkosz *et al.*, 1990)

County	Soil drainage classes		
	Well	Moderately well and somewhat poorly	Poorly and very poorly
Sandstone-shale colluvium			
Centre	36	27	24
Huntingdon	36	40	15
Juniata and Mifflin	51	26	15
Limestone colluvium			
Centre	11	2	–
Huntingdon	6	3	–
Juniata and Mifflin	7	1	–

Thus the soils and landscapes of the Ridge and Valley of central Pennsylvania 'show a distinctive imprint of the climatically-controlled processes that have acted with varying intensity over time in this area' (Ciolkosz *et al.*, 1990, p.258). Soils on the ridge and shoulder slopes have developed in sandstone colluvium and are of late Wisconsin age. Material from the ridge-tops has moved to the footslopes truncating material on the backslopes in the process. Adjacent to the footslope area are soils that have developed from residuum that accumulated when calcium carbonate was leached from the limestone. Superimposed on this activity has been a not inconsiderable aeolian input.

The Appalachian Piedmont and Coastal Plain are adjacent provinces with similar long-term climatic histories. From west to east across the Piedmont and from east to west across the Coastal Plain, soil characteristics indicate that soils in both provinces reach maximum maturity near the Fall Zone (Figure 5.5). Ultisols are the dominant soil order but vary widely from the least-developed Ultisols of the western Piedmont and eastern Coastal Plain to the thick Paleudults of the inner Coastal Plain near the Fall Zone (Markewich *et al.*, 1990). Erosional stability of the geomorphic surfaces seems to determine the distribution of soils. The Piedmont is composed of deformed and metamorphosed metasedimentary, metaigneous and metavolcanic rocks of Late Precambrian to early Palaeozoic age intruded by mafic dikes and granitic plutons. In terms of morphology it is an eroded sloping plain with altitudes ranging from 300 to 650 m at the Blue Ridge Front to 30 to 45 m in river channels at the Fall Zone. Occasional residuals rise above the plain (e.g. Stone Mountain, Georgia, Sugarloaf, Maryland). Slopes of interfluves appear accordant and relative relief appears constant over large areas. Based on the work of Markewich and Pavich (Markewich, 1985; Markewich *et al.*, 1986, 1987, 1989; Pavich, 1985, 1986, 1989; Pavich *et al.*, 1985, 1989) it is believed that soils on the Piedmont uplands have developed within the Quaternary and that the Outer Piedmont is a Pleistocene landscape with isolated remnants of Miocene to Pliocene age regolith.

The Coastal Plain is a seaward-sloping surface underlain by Late Cretaceous to Holocene fluvial and marine sediments. Thick, mature soils (Paleudults) are common near the Fall Zone and in the central part of the Coastal Plain 2 to 5 m thick Hapladults and Paleudults have developed (Markewich *et al.*, 1990). Constructional landforms that are Pliocene and younger in age have been only minimally eroded and soils associated with these landforms have been used to establish chronosequences of landform development and to determine a minimum age of the specific landforms (Markewich *et al.*, 1987; Daniels *et al.*, 1978). A number of systematic changes occur across the Coastal Plain from west to east. There is an increase in the number, areal extent and degree of preservation of constructional landforms and a decrease in depth of

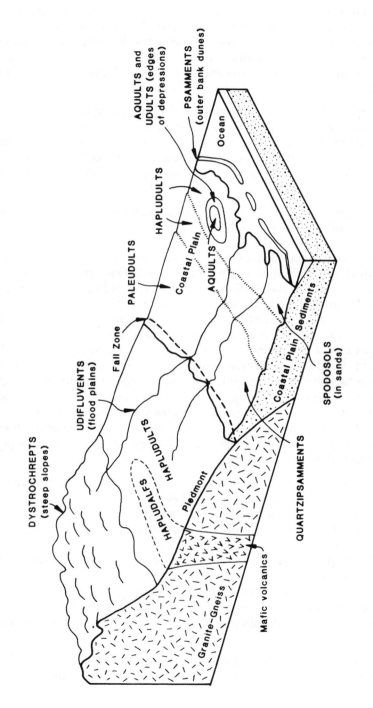

Figure 5.5 Soil-landscape relationships in the Appalachian Piedmont and Coastal Plain areas of eastern United States (after Markewich *et al.*, 1990).

weathering from 20 m on the oldest units to l m on the youngest. The degree of soil development decreases and there is an increase in the areal extent of organic-rich Spodosols and Ultisols with spodic horizons. There is an increase in number, depth and thickness of humates with an increase in age from Pleistocene to Pliocene. Markewich *et al.* (1990) assert that:

> comparison of soil properties on adjacent landscapes underlain by different parent materials is an extremely useful approach to understanding landscape processes and evolution in the eastern United States, and should lead to better understanding of the more specific effects of climatic change and tectonic processes on landscape evolution (Markewich *et al.*, 1990, p. 445).

Weathering and pedogenic processes appear to increase the thresholds or inertia of a landscape and increase its resistance to rapid response to external force.

The Swan Coastal Plain of Western Australia is another area where soil relationships have been examined in detail (McArthur and Bettenay, 1960). The coastal plain is formed almost entirely of fluviatile and aeolian depositional material and consists of a series of landform units subparallel to the present coastline. The Pinjarra Plain is the main landform unit and is bounded on the west by a series of coastal sand dunes. The pattern of the plain is a series of coalescing piedmonts which can be subdivided into distinct depositional systems with their age relationships established stratigraphically. The names given to the systems – Coolup, Wellesley, Boyanup, Blythwood, Belhus, Dardanup, Vasse and Pyrton – are those of the major soil series and they range in age from approximately 400 000 years BP (Coolup) to the presently forming Pyrton system. As a summary the following are some of the temporal soil trends that have been identified:

1. Colour changes from dark-brown through brown, red and yellow to grey with yellow-brown mottles because of iron segregation.
2. Differentiation in texture and colour becomes more marked.
3. Less stable primary minerals break down and the ratio of clay to silt increases.
4. The proportion of kaolinite to illite increases and gibbsite forms in the final stages.
5. The percentage of saturation of the exchange complex decreases.
6. The ratio of calcium to magnesium in the exchange complex decreases, calcium dominates in the younger soils and magnesium in the older.

The separation between the various elements of this sequence is not always clear cut altitudinally which is far from the case in many coastal plains such as those of the south-east United States. In this area the age sequences are also different in that the unconsolidated fluvial, marine and aeolian deposits range in age from Pleistocene to Cretaceous. The surfaces form a prominent stepped sequence with well-marked intervening scarps, and in a series of articles, Daniels, Gamble and their associates have established the soil differences between them (e.g. Daniels *et al.*, 1970; Daniels *et al.*, 1971; Daniels and Gamble, 1978). The middle and upper surfaces are composed essentially of fluvial materials although there is a small marine component, whereas the lower surfaces are mainly marine in origin. Because of the extreme age of some of these surfaces and the similar material composition, the same soils or group of soils can occur on more than one surface. But, nevertheless, distinct decreasing trends in features such as solum thickness, depth to watertable, depth to and thickness of plinthite and percentage gibbsite content, can be identified from older to younger surfaces.

CONCLUSIONS

The analysis and identification of erosion surfaces was common geomorphological practice in the 1950s and 1960s. But, as most of the studies were based almost entirely on morphology, many of the results were inconclusive. The analysis of soils has added a new dimension to studies of erosion surfaces; similarly, erosion surfaces provide a chronological framework within which to study soils. But soil properties and time cannot be directly related to the age of the surface. The surface may be appreciably older than the sediments and soils which mantle its surface. Also, it is the time a particular process has been active that is important and not necessarily the total time that soil material has been exposed to weathering. Time zero might not be the same for two measured properties. But, although a surface may have many kinds of soils these soils will have a common degree of soil development.

6

Soils on floodplains

Floodplains and floodplain soils are of special interest to many earth scientists. Floodplains occupy a significant proportion of the Earth's surface, approximately 2% of Africa, 3% of South America and a greater proportion of tropical Asia. Alluvial soils are generally associated with river floodplains but alluvial soils are difficult to define satisfactorily because they may be developed on fluvial, lacustrine or marine deposits. Thus alluvial soils occur on coastal plains, deltas, river terraces and alluvial fans as well as river floodplains. Soils developed on marine sediments have very distinctive characteristics and are examined in Chapter 8. Soils on river terraces are often called alluvial but many river terraces are comparatively old and soils formed on them are often well-developed and completely different to soils on recent fluvial sediments. Such soils and landforms are examined in Chapter 7. The geomorphological development of floodplains has a tremendous effect on soil development. The temporal development of soil types follows a chronology imposed by the deposition of floodplain sediments, while such soils are strongly affected by the variety of landforms, relief, texture and moisture status of floodplains. These relationships are now examined.

CHARACTERISTICS OF ALLUVIAL ENVIRONMENTS

River-formed landscapes are reasonably similar the world over in terms of both landforms and processes. This does not imply that soils from diverse river areas are identical but that the natural landforms and soil patterns are similar. Soil materials and drainage conditions show sharp boundaries related to major landform types (Figure 6.1). Natural levees, consisting of sandy sediments, occur adjacent to present or former river beds. Levees form the highest parts of the landscape and, consequently,

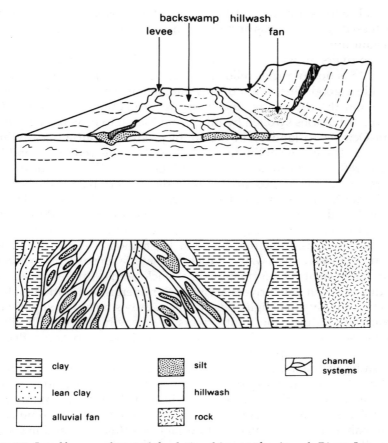

backswamp　hillwash
levee　　　　　　　fan

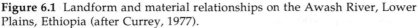

	clay		silt		channel systems
	lean clay		hillwash	
	alluvial fan		rock	

Figure 6.1 Landform and material relationships on the Awash River, Lower Plains, Ethiopia (after Currey, 1977).

are freer drained. Behind the levees lower-lying areas occur formed of heavier-textured and impermeable silts and clays. Accumulation of organic matter in these backswamps is common.

Levees and basins are universally important and this means that soil maps of rivers such as the Rhine, Mississippi, Indus, Tigris and Euphrates look remarkably similar. There are, neverthless, major differences between the alluvial soils of these river basins, and these differences tend to reflect the larger geomorphological and climatic systems of which the rivers are a part. A major distinction has been made between alluvial soils of tropical and temperate areas (Edelman and Van der Voorde, 1963). In temperate areas floodplain soils are generally rich in mineral nutrients, partly due to the effect of glaciations providing powdered fresh minerals to be transported and deposited by rivers. Many tropical and subtropical rivers have their headwaters in deeply-weathered igneous areas and their deposits consist of quartz and other resistant minerals.

Large rivers, in whatever environment, will be governed by the complex interaction of rocks, relief and climate. Rivers in the Amazon Basin provide a good illustration of this interaction. The size of the Amazon Basin and the climatic regimes it spans means there is no single flood season; tributaries swollen by rains from March to July in the north and from October to January in the south produce a pulsating rhythm of flooding. The colour of the rivers, yellowish-white, red-black, and blue-green, reflect Amazonian geology (Sperling, 1973). The major white rivers (except the Branco) are found in the western part of the basin and rise in the Andes. The black rivers, such as the River Negro, tend to drain the highlands of northwestern Brazil and Venezuela and flow south to the main river. The blue-green rivers, such as the Xingu and Tapayos, flow northwards from the Brazilian Highlands. The black and blue-green rivers have clear water, with little material in suspension and flow from the ancient rock formations that enclose the Amazon Basin to the north and south. Such rivers are poor in minerals and nutrients. Rivers draining the Guyana Highlands are chemically similar but slightly more turbid. In contrast, the white rivers are heavily laden with silt and soluble soil nutrients. Alluvial soils of the lower Amazon reflect this complexity of river processes.

In India distinctions can be made between the arid-zone floodplains of Punjab, Sind and Rajasthan; the intermediate conditions of Uttar Pradesh; and the wet monsoon alluvium of the Ganges–Brahmaputra floodplains. In the arid zone, alkaline and saline soils are common, the savannah areas possess neutral to weakly alkaline soils and in the high rainfall areas acid gleyed soils occur. Calcimorphic soils can occur with a rainfall of about 600–750 cm, which with freer drainage will produce ferruginous soils. The reaction in the alluvial soils is often less acid than in non-alluvial soils of corresponding climates and is usually in the pH range of 6 to 8. In Bangladesh pH is 6–6.5 on older alluvium and 7–8.5 on the newer alluvium of the Ganges, whereas on currently inundated areas of the Brahmaputra, it is 5.5 (Islam, 1966).

A contrast can also be made, especially in temperate areas, between floodplains in uplands and lowlands. In upland areas coarse channel sediments occupy much of the valley floor and forms a combination of relict bar forms and migratory channel patterns with a small amount of fine overbank sediment on top. Floodplains, in such areas, are built up largely of coarse gravel in lateral and medial bars with fine sediment accumulation in cutoffs and inner bar margins.

Floodplain soils are usually thought of as being young or undeveloped. But not all such soils are undeveloped in the strict pedological sense, since some changes that resemble soil formation may have taken place. Thus, some soils of alluvial swamps have distinct dark-coloured topsoils that have the characteristics of A1 horizons. Although major differences between alluvial areas exist, basic interrelationships between minerals and geomorphological and pedological processes provide a unifying theme within which to view the evolution of soils.

SEDIMENTARY AND PEDOLOGIC CHARACTERISTICS

Floodplain soils exhibit characteristics of both sediment transport and deposition, and soil formation. Soil formation and sedimentation overlap and sites protected from erosion and undergoing slight vertical accretion develop distinct horizons. Thus soil profiles should be examined with relation to sedimentary environments. Sediment stratification is common and has a considerable effect on the development of pedogenic properties. Ruhe (1975) has drawn attention to this problem in the floodplain of the Missouri River. Within the channel belt, soils have light-coloured horizons and their subsoils are stratified and show little pedogenic weathering. Soils in the meander belt have thicker dark-coloured surface horizons and their subsoils show pedogenic structures. In the Wolf Creek floodplain in Iowa, ridge soils have little profile development but depression soils have clayey B horizons with strong subangular blocky structure. Both soils show sedimentary stratification, but whereas ridge soils have little vertical organization of clay, depression soils have good vertical organization.

Failure to take the effects of stratification into consideration leads to an overestimation of the rates of pedogenic processes (Arnold, 1968; Brewer, 1968; Oertel, 1968), and it is, therefore, important to be able to separate sedimentological from pedological characteristics. Several approaches to this problem have been suggested; the main principle employed being the disunity of depth functions of soil properties (Foss and Rust, 1962; Sleeman, 1964; Oertel and Giles, 1966). The method adopted by Raad and Protz (1971) is a variant of this theme. They identified sediment stratification by establishing statistically significant changes with depth in the values of the ratios of percentage sand to percentage silt. This enabled prominent stratifications to be identified, clearly important in determining the rate and type of soil formation.

An interesting addition to this work has been suggested by Green (1974) based on the sedimentological work of Moss (1962, 1963, 1972), involving long (p) and medium (q) dimensions of grains. It seems possible to separate *in situ* soils from transported sediments by plotting the elongation function p/q against p for different grain sizes. Elongation function curves for an *in situ* granite soil showed that its quartz grains became increasingly equant with increasing size. The coarser particles of the soils were less equant and more elongated with increasing grain size. This also seems to be characteristic of bed-load sediments. On this basis it is possible to distinguish transported from *in situ* parent materials. Other indices have been developed by Bilzi and Ciolkosz (1977) and Meixner and Singer (1981). However, the pattern of textural layering is usually complicated, random and unpredictable (Karageorgis *et al.*, 1984). A succession of flood deposits may remain visible even when soil-forming processes have been active for some time. Old soils can retain evidence of the depositional environment even when no visible trace of lamination remains.

Stratification in floodplain soils becomes obliterated by the combined processes of ripening and homogenization (Hoeksema, 1953). Ripening involves the drainage and evaporation of excess water, and the development of drying cracks which allows air to penetrate the soil and may lead to loss of part of the organic matter by oxidation. Homogenization refers to the elimination of depositional layering by biological agents such as roots, worms, termites and various micro-organisms. In later stages, clay translocation continues the process, and the presence of peds is an indication that soil formation is occurring. Crumblike and fine subangular blocky peds are common in the root zone but their development is restricted where the water-table is permanently high. Subsoil peds, often attributed to wetting and drying stresses, are restricted in soils with a high water-table. Most floodplain soils show characteristics associated with waterlogging and flooding.

Many of these features have been examined in the floodplain soils of the River Severn, England (Hayward, 1985). The better developed soils exhibit trends in which soil structure, organic carbon and pyrophosphate-extractable iron decline with depth. Subangular blocky structure weakens with depth in soils not affected by additions of surface material. Hayward develops three assumptions related to the rising alluvial profile. These are:

1. Floodplain materials go through a succession described as sediment-topsoil-subsoil-deposit. In the topsoil phase the soil inherits some properties from the original sediment. Then the material is isolated from the source of material and such properties should decline. In later stages weathering continues and the soil develops.
2. The relationship of the rising alluvial profile with respect to the water-table. Although the water-table will tend to rise as the channel rises during aggradation, later additions of sediment will tend to raise the channel higher than the water-table. Thus gleying will change to oxidative weathering. Decalcification and clay translocation will also occur.
3. Relationship between sediment rates and soil profile development. The rising alluvial surface will tend to follow the lower curve in

Figure 6.2. This is Hayward's (1985) interpretation of a curve of Nanson (1980). Intensity of soil forming processes will tend to follow the upper curve based on Birkeland (1974, 1985). As the floodplain surface approaches a hypothetical level of aggradation the intensity of soil-forming processes should increase. Hayward (1985) suggests that the development of floodplain soils in the manner suggested would result in greater horizon development at the top of the alluvial sequence.

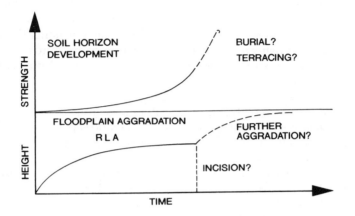

Figure 6.2 Model of soil development on an aggrading floodplain (from Hayward, 1985).

WATERLOGGING AND THE EFFECTS OF FLOODING

Many of the characteristics of floodplain soils are due to waterlogging. Gleying is common to most floodplain soils and, in immature soils, it is associated with laminae produced by deposition. In more mature soils it is related to the developing soil structures. Surfaces of peds and linings of pores become reduced under waterlogged conditions, producing the grey colour that contrasts with the mottled unreduced ped interiors. Soils on low terraces depend on rainfall to induce seasonal wetness, whereas on floodplains these effects are superimposed on the effect of the fluctuating water-table. Soil horizons are usually of three types:

1. A partly oxidized A horizon with a high organic content;
2. A mottled zone within which oxidation and reduction alternate; and
3. A permanently reduced bluish-green zone.

Under reducing conditions certain elements, especially iron and manganese, become increasingly mobile. The dissolution of iron is encouraged by the formation of ferrous-organic complexes by chelating agents from fresh organic debris. Some of the ferrous iron diffuses into the peds where it may be oxidized when the water-table falls. Iron concretions may occur in zones of high pH or high redox potential and are likely to survive changed hydrologic conditions such as lowering of the water-table.

In the mottled zone, iron and manganese are deposited as rusty mottles, or streaks if diffusion of oxygen into soil aggregates is slow (Ponnamperuma, 1972), but if diffusion is fast they are deposited as concretions. The concretions are rarely pure oxides and may contain small amounts of zinc, copper, nickel and cobalt. The grey colour in gleys is essentially due to the presence of organic matter closely associated with clay particles (Bloomfield, 1973). The permanently water-logged zone is bluish-green because of ferrous compounds. The blue colour may be the result of vivianite (ferrous phosphate) but hydrated magnetite, pyrite, marcasite, siderite and ferrous silicates also occur. Spots of black are likely to be ferrous sulphide.

It has been suggested that, in groundwater gleys, iron will move towards the major cracks and pores around which iron oxide separations will form (Brammer, 1964; Schuylenborgh, 1973). Some of the ferrous iron will diffuse into the soil peds. As mentioned earlier iron concretions occur in zones of high pH or Eh and are likely to survive changed hydrological circumstances, such as lowering of the water-table, because they can be remobilized only by chelates. Manganese compounds are more easily reduced than ferric compounds and will concentrate in areas higher up the profile – laboratory experiments support such field results (Veneman *et al.*, 1976; Vepraskas and Bouma, 1976). Observed mottle patterns were matched with moisture regime and redox fluctuations. Such periods of saturation lead to manganiferous mottles being formed. Long periods of saturation lead to the removal of manganese from the profile.

Ferrolysis is the term used for the hydromorphic soil-forming process in which a soil's cation exchange capacity is destroyed due to exchange reactions involving iron in seasonally alternating cycles of reduction and oxidation (Brinkman, 1970). The process described by Brinkman was based on extensive work in Bangladesh and involves a sequence of aerobic and anaerobic cycles, with oxidation of organic matter providing the energy to power the ferrolysis cycle. Under anaerobic conditions, free iron is reduced with the formation of hydroxyl ions, the ferrous iron displaces exchangeable cations and the displaced cations are leached. Under aerobic conditions, ferrous iron is oxidized, producing ferric hydroxide and hydrogen ions. The hydrogen ions displace the exchangeable ferrous iron and corrode the octahedral layers of clay minerals at their edges. Cations are leached and part of the clay lattice is destroyed during each cycle. Thus, a seasonally wet soil can develop into a grey, unstable, silty or sandy soil with low clay content and low cation exchange capacity.

Periodic flooding causes a variety of other soil features. Many seasonally flooded soils possess coatings of clay on the soil peds (Brammer, 1964, 1966, 1971). These coatings are especially common in the seasonally flooded soils of Bangladesh where they are typically continuous, thick and grey in colour. They occur in the young floodplain soils of the Brahmaputra and Ganges alluvium as well as in some terrace soils. They have also been observed in the Guadalquivir floodplain soils of Spain, in northern Malaysia and in the Indus plain. These coatings may have been mistaken in the past for clay skins, cutans or argillans and therefore overlooked.

Flood coatings differ from argillans in the rapidity with which they develop and, to a certain extent, in their composition. They consist, not only of fine clay, but also of silt and humus and, even in young floodplain soils, may be 0.5 mm thick. Their colour suggests that they consist of material from the soil itself and are not created by sediment brought in by floodwater. On the Ganges floodplain, where sediments in the floodwater tend to be calcareous, the coatings are non-calcareous in soils with non-calcareous topsoils. The coatings only occur in soils that are flooded for part of the year and are best developed in deeply-flooded soils, and least in floodplain ridge soils only intermittently flooded. Therefore, the formation of the coatings seems to be directly related to flooding.

After continuous submersion for about two weeks, the top few inches of the soil become strongly reduced. Subsoil layers are, apparently, unaffected and remain oxidized. Topsoils that have undergone these seasonal alternations of reduction and oxidation for a long period will have lost most of their free iron by vertical leaching and will be in an easily dispersed condition. But it is still not clear how the dispersed topsoil moves to lower layers. It may occur at the start of the flood season when the topsoil will be saturated before the lower layers and dispersed soil may flow by gravity down subsoil cracks and pores. It could also occur after flooding ceases when rains may wash soil down cracks opened up in the drying soil. But these conditions occur in a variety of non-flooded soils and do not seem to result in the formation of coatings. This re-emphasizes the relationship between coatings and flooding. The material in the coatings is finer than the adjacent soil mass, and the coatings are often banded, indicating successive flows, but the clay appears to be non-oriented. It is considered by Brammer (1971) that the coatings move as thin mudflows possibly under hydraulic pressure from the weight of floodwater resting on the soil surface. Whatever their origin, these coatings appear to be specifically related to seasonal flooding. Hayward and Fenwick (1983) have suggested that micromorphological evidence may suggest at what stage any observed void or grain coatings are formed.

FLOODPLAIN LANDFORMS

Many of the characteristics of floodplain soils are related to the distinctive nature of the landforms. A comprehensive review of floodplain geomorphology has been provided by Lewin (1978) and only the major elements are discussed here. The main landform types are illustrated diagrammatically in Figure 6.3. and although this is a composite diagram, each type can be frequently matched with actual examples. The complexity of cut-off processes on the River Mississippi has been comprehensively described by Russell (1967), whilst crevasse splay development is easily seen on the Seyhan River, Turkey (Russell, 1967).

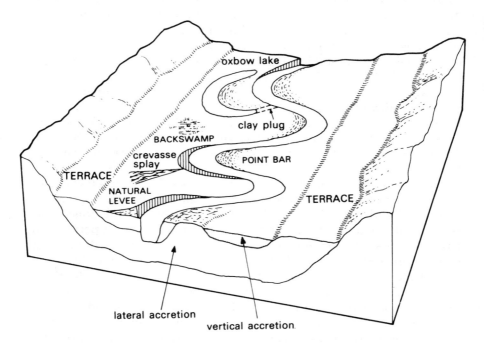

Figure 6.3 Landforms and deposits of a typical floodplain: Va, vertical accretion; La, lateral accretion (adapted from Vanoni, 1971; Gregory and Walling, 1973).

Each landform type is associated with a slightly different type of sediment and, therefore, the basic soil materials differ quite considerably. Point bars are built up of layers of coarse gravel with smaller amounts of sand, and levees are generally formed from the coarser materials carried by turbulent flow. During river flows in excess of the bankfull stage, the water extends beyond the levees, becomes less turbulent and begins to flow as sheets. Bed load is deposited close to the channels whereas the suspended load, with the finer silts and clays, accumulates in the backswamp areas. The rate of sediment accumulation depends on the size and quantity of sediment and the frequency of flooding. Coarse material settles rapidly but finer clays take longer. In areas where flooding is infrequent, lateral accretion and channel deposition are most significant in the formation of floodplains. In the United States of America, Leopold and Wolman ((1957) have estimated that 80–90% of floodplain deposits have been formed in this way and only 10–20% by the process of overbank deposition. However, in other areas such as the humid tropics, where fine-grained material is dominant and where flooding is frequent, overbank deposits assume greater importance. In Papua New Guinea, especially along the Fly and Strickland Rivers, Blake and Ollier (1971) have shown that backswamps and overbank deposits are dominant.

Landforms can be examined on three scales (Lewin, 1978):

1. Of the floodplain and channel pattern;
2. Of landforms of the magnitude of the channel; and
3. Of small-scale features such as ripples and dunes in the bed of the river.

Soil-landform relationships are developed best on the first two scales. The intermediate scale forms are the levees, basins, cut-offs and crevasse splays portrayed in Figure 6.3. Such features are not present in all floodplains but there are usually consistent relationships (Table 6.1). Thus, in the area of the Missouri floodplain examined by Schmudde (1963) levees are rare and much of the floodplain has been reworked recently. This contrasts with the lower Mississippi where levees are prominent.

In general, levees are the most conspicuous floodplain landforms, but vary considerably in size. The Columbia River possesses levees 50 m wide and 2.5 m high whereas those in the lower Saskatchewan River are 1 km wide and 4 m high (Smith, 1983). The levee-building process has been well-documented (e.g. Alexander and Prior, 1971; Kesel *et al.*, 1974; Ritchie *et al.*, 1975; Ritter *et al.*, 1973; Weidner, 1981). Levee deposits are laminated fine sand and silt with occasional organic lenses.

Crevasse splays exhibit good relationships between sedimentation, texture and landforms (Knight, 1975). They form when an overbank flow cuts a channel through a levee and then deposits sediment as a lobate sheet on the backswamp area. Most splays exhibit a pattern of a coarsening and then a fining upward. As the splay develops, a basal unit of coarse silt and very fine sand is deposited beyond the prograding front of the sand-sheet onto peat, marsh or lacustrine deposits. Waterlogged vegetal litter may be buried between sand layers. In the Columbia Valley, splays vary from 0.3 km to 1 km with deposits 2–3 m thick (Smith, 1983).

Slack water deposits need not form only as a result of overbank processes. Where rivers are confined by bedrock, fine-grained sediments

Table 6.1 Landforms associated with some river floodplains (modified from Lewin, 1978)

River	Reference	Floodplain width (m)	Braid bars	Scroll bars	Abandoned channels	Levees	Flood basins
Brahmaputra	Coleman, 1969	13 000	A	P	P	A	A
Fly River, Papua New Guinea	Blake and Ollier, 1971	3000–16 000	D	A	P	D	A
Gila, U.S.A.	Burkham, 1972	300–15 000	P	P	D	P	D
Glomma, Norway	Nordseth, 1973	15 000	A	D	D	D	D
Little Mississippi, North Dakota, U.S.A.	Everitt, 1978	880	R	A	P	D	D
Missouri, U.S.A.	Schmudde (1963)	3540	R	A	P	R	P
Mississippi, U.S.A.	Fisk, 1952	40 000–200 000	P	A	A	A	A
Rheidol, Wales	Lewin, 1978	880	P	A	P	D	D
Spey, Scotland	Lewin and Weir, 1977	16 000	A	R	P	D	D

Note: A = Abundant; P = Present; R = Rare; D = Does not occur.

occur from suspension and become the dominant mode of floodplain development. Good examples of such slack water deposition are found in the rivers of Northern Territory, Australia (Baker *et al.*, 1983). Slack-water deposition also occurs in the mouths of tributaries (Stewart and Bodhaine, 1961; Moss and Kochel, 1978; Koechel and Baker, 1982).

Numerous bar types can occur and the terminology is somewhat confusing (Miall, 1977). However bars are usually of three main types: longitudinal, linguoid (transverse) or point (lateral) bars. A succession of point bars and swales forms a meander scroll complex. Floodplains often contain various types of lakes. Oxbow lakes in abandoned meanders (Weihaupt, 1977), embankment lakes in depressions dammed by levees and serpentine lakes in sinuous abandoned channels. Lake sediments consist of laminated silts and clays and, in combination with aquatic vegetation, produce highly organic distinctive soils.

Specific associations exist between levees, crevasse splays and soils. In the lower course of the River Rhine in the Netherlands, levees have a soil morphology corresponding to effective drainage (Havinga and Opt'Hof, 1983). However, in some places, due to the low position of the top of the levee with respect to the surface in the adjoining basin, soil profiles exhibit hydromorphic features. Soils on splays are more complicated (Havinga, 1969).

Many of the general principles just described can be illustrated by taking examples from the major river systems of the world. One of the best documented is the Lower Indus Plain. The lower valley of the River Indus possesses an intricate pattern of channels, levees, cut-offs and sand bars. Occasionally, exceptionally high summer floods have allowed the river to shift its channel and establish a new route, but normally the annual floods simply overtop the levees causing flooding but no major course changes. These processes have created a number of distinctive landforms together with their associated deposits (Figure 6.4).

The present course of the Indus is contained by artificial banks and a large proportion of its flow is diverted into canals to feed one of the largest irrigation systems in the world. Alluvial deposition has, therefore, been considerably curtailed, but recent former courses of the river can be traced by the presence of distinctive landforms such as meander scars, meander scrolls and deserted channels. Two distinctive elements can be recognized: the broad floodplain of the major river was able to contain much of the annual floods restricting the development of levees (large and numerous meander scars are its main features); a second type of floodplain was formed by minor branches of the river, here the floodplains are narrow, with bars and meanders restricted to a single channel. Annual floods easily overtopped the levees forming extensive cover floodplains. These major landforms and their associated deposits are listed in Table 6.2.

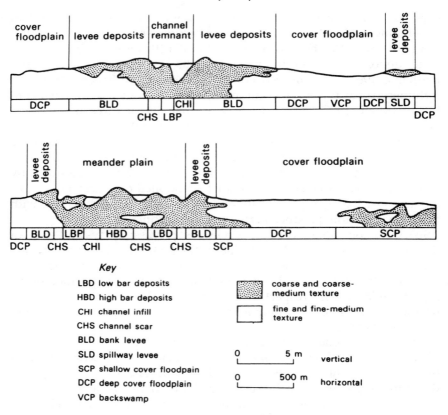

Figure 6.4 Relationships between floodplain landforms and materials in the Indus Plain (from Holmes and Western, 1969). (a) A channel remnant surrounded by cover floodplains is separated by extensive levee deposits. (b) A meander plain is located near an earlier buried meander plain, now a shallow cover floodplain. The texture pattern in (b) is considerably more complicated than that shown in (a).

TEXTURE OF THE MATERIALS

A detailed textural investigation of the various deposits of the Indus Valley has been undertaken by Holmes and Western (1969). They devised a symbolic nomenclature to distinguish both texture, comprising coarse, medium-coarse, medium-fine and fine, and depth relationships, i.e. whether coarse-textured material overlay fine or vice versa. The nomenclature used is shown in Table 6.3. An examination of the detailed textural classes shows that individual landform associations tend to be dominated by one or two classes. Thus, levees are dominated by class B, backswamp deposits by class Dv and shallow cover floodplains by class C (Table 6.4). This produces the general associations portrayed in

Table 6.5. There is clearly a general relationship between texture patterns and factors such as elevation, relief, waterlogging, salinity, permeability and drainage. This generates a complex textural map but always the relationship between landforms and material is evident.

Table 6.2 Geomorphological components of the Indus floodplain (adapted from Holmes and Western, 1969)

Geomorphological subregions	Landform association	Deposits
meander floodplain	meander plain channel remnant active floodplain	high bar low bar channel-scar channel infill
levees	levee forms	bank levee spillway levee flood levee anthropic levee
levee floodplain	levee floodplain	shallow cover deep cover backswamp

Table 6.3 Nomenclature for describing sediment characteristics in the Indus Valley (from Holmes and Western 1969)

Nomenclature	Characteristic texture
AL	dominantly coarse sand and loamy sand
A	mixture of sands and loams
Bd	sands and loams overlie silts and clays at depth > 100 cm
B	sands and loams overlie silts and clays at depths between 50 and 100 cm
Bs	sands and loams overlie silts and clays at depths < 50 cm
Xlm	complex but basically sands and loams
X	complex sequences of fine and coarse textures
Xhv	complex but basically silts and clays
Cs	silts and clays overlie sands and loams at depths < 50 cm
C	silts and clays overlie sands and loams at depths between 50 and 100 cm
Cd	silts and clays overlie sands and loams at depths > 100 cm
CdV	upper horizons dominated by clay
D	almost entirely silts and clays
Dv	clays probably dominant

Similar relationships have been shown by Ruhe (1975) along the River Missouri (Figure 6.5). Sites 4 and 8 are on natural levees and are dominated by sand and silt in the upper parts of the profile. Sites 5, 6 and 7, in the intervening basins, are dominated in the upper parts by silt and clay with

Table 6.4 Relationship between texture and landforms in the Indus Valley (from Holmes and Western, 1969)

Landform association	AL	A	Bd	B	Bs	Xlm	X	Xhr	Cs	C	Cd	Cdv	D	Dr	Total sample
low bar deposits	41	27.5	0.5	4	–	4	2	–	14.5	6.5	–	–	–	–	200
channel scar deposits	26.7	26.7	–	6.7	–	6.7	33.3	–	–	–	–	–	–	–	15
channel infill deposits	0.7	–	–	–	0.7	–	1.3	5.7	2.5	24.8	9.6	–	35.0	19.7	157
levee deposits	10.2	21.3	5.5	33.1	7.1	12.7	21.6	–	0.8	–	–	–	–	0.8	127
shallow cover floodplain	0.4	1.8	–	0.4	–	0.4	10.2	11.1	5.8	53.3	8.9	–	6.7	0.9	225
deep cover floodplain	–	–	0.5	1.0	1.0	–	–	0.5	–	0.5	8.3	–	80.7	7.3	192
backswamp deposits	–	–	–	0.6	–	–	0.3	1.5	–	2.4	0.3	5.4	6.3	83.3	336

Percentages sites per textural group

a reversal occurring in the lower parts of their profiles. The lower sediments are channel accretion deposits whilst the upper sediments are the result of vertical accretion. Many other examples of similar relationships can be cited, such as those of the Tigris and Euphrates Valleys (Buringh and Edelman, 1955) and of the Llanos Orientales of Colombia (Goosen, 1972).

Table 6.5 Characteristic textures of landform types in the Indus Valley (from Holmes and Western, 1969)

Landform association	Characteristic texture
high bar deposits	AL
low bar deposits	A, AL, Cs, Xlm
channel scar deposits	A, AL, Cs, Xlm
channel infill deposits	D, Dv, Cd, CdV
bank levee deposits	D, AL, Bd, B, Xlm
spillway levee deposits	A, Bd, B, Xlm
flood levee deposits	Al, A, Bd, B, Bs, Xlm
anthropic levee deposits	A, Bd, B, Bs
shallow cover floodplain	CdV, Cd, Dv, D
backswamp deposits	Dv, CdV

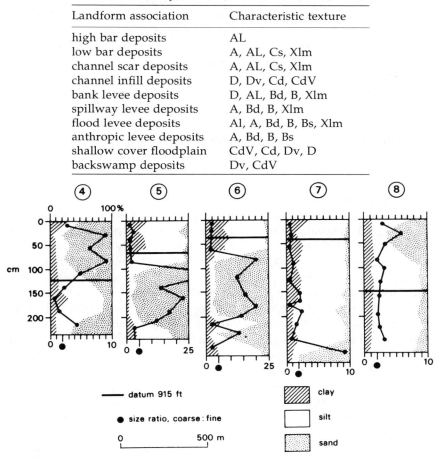

Figure 6.5 Soil changes with position and landforms on the Missouri floodplain (after Ruhe, 1975).

CLASSIFICATION OF ALLUVIAL SOILS

Alluvial soils, because they are related to specific landforms (e.g. flood-plains) that occur in a variety of climatic regimes, have always been outside the main classification schemes. Early schemes, such as those of Dokuchaev (1879, 1893) and Sibirtzev (1895), classified soils as normal (zonal), transitional or abnormal. Alluvial soils were placed in the last category (Afanasiev, 1927a, 1927b). The terms transitional and abnormal were eventually replaced with intrazonal and azonal (Sibirtzev, 1901), azonal soils being soils which did not reflect the characteristics of any particular climatic zone, usually because they were too young. Thus, markedly different soils can be classed as alluvial soils and such soils may or may not have some of the local zonal influences affecting their characteristics. In more recent work on soil classification to emerge from the USSR more emphasis is given to the moisture characteristics of the soil (Kovda *et al.*, 1967; Rozov and Ivanova, 1967). This has especial significance for alluvial soils. Moisture characteristics are described as automorphic (aerobic), weakly hydromorphic (weakly anaerobic), semi-hydromorphic (moderately anaerobic) and hydromorphic (anaerobic). Duchaufour (1982) uses similar terminology to subdivide alluvial soils namely, slightly hydromorphic and hydromorphic. He divides slightly hydromorphic into low humic alluvial soils and humic alluvial soils, and hydromorphic alluvial soils into low humic alluvial soils with gley and peaty alluvial soils.

The standard ABC model of profile development has been shown to be inappropriate for floodplain alluvium-derived soils (Riecken and Poetsch, 1960). Soil profiles containing different soil stratigraphic layers complicate the classification process (Finkl, 1980). Systems such as that of FitzPatrick (1980) which does not rely on the unity of the soil profile, or that of Marinkovic (1964), which is based on factors of topsoil texture and depth, may be more applicable to floodplain soils. The standard ABC model assumes that the organic matter in the A horizon has formed in place after the parent material was deposited. In floodplain soils some organic matter will have been carried in the alluvium, also, new C material will have been added at the surface. Thus, the subsoil forms in a former A horizon and the present A horizons of young alluvial soils are formed in new superficial C material. The present B horizon will form in a former A horizon. Therefore the genetic pathway is not C transformed to A or B directly but a new superficial C into a new A and a former A into a B. A former B horizon becomes a substratum not to be confused with C horizons.

The specific sequence will depend on the rate of accumulation of material in relation to the rate of soil formation. A horizons are easily destroyed on rapidly accumulating sites, but where the intervals between deposition are longer the A horizons are deeper and more distinct. These characteristics are well-displayed in soils of the lower Rhine floodplain (Havinga and Opt'Hof, 1983). It is possible to identify, in lower Rhine floodplain basins, at least three and often four phases of fluvial deposition separated by fossil vegetation layers. The same sequence is not found in soils on the levees and it is not easy to correlate such soils with those in the basins.

This problem is seen in the system of Kubiena (1953). Soils are classified as (A)C, AC, A(B), A(B)C, ABC or B/ABC depending on the presence and relative importance of the respective horizons. As noted above many floodplain soils have (A)C profiles. But, some soils derived from alluvium are no longer subject to flooding and have A(B)C or even ABC profiles.

The difficulty of classifying alluvial soils is demonstrated in the United States Department of Agriculture scheme (Soil Survey Staff, 1975). Alluvial soils can be classified either within the order of Entisols or Inceptisols. Entisols are characterized by an absence of pedogenic horizons and a dominance of mineral soil materials, and there is no evidence of translocation or alteration of minerals. Most floodplain soils developed on recent alluvium would fall into this category and into the suborder of fluvents. Fluvents are mainly brownish or reddish soils which have formed on recent floodplain sediments that are frequently flooded and in which stratification is normal. However, waterlogged soils, with the mottled, greyish or bluish horizons discussed earlier, are aquents and some sandy soils on levees are psamments.

Some alluvial soils are classified in the Inceptisol order. These soils are more developed with one or more pedogenic horizons but still with little accumulation of translocated material apart from carbonates or amorphous silica. Most alluvial soils would be included in the aquepts suborder. Their natural drainage is poor and the water-table is close to the surface at some time of the year. They usually have a grey to black surface horizon and a mottled grey subsurface horizon that begins at a depth of less than 50 cm. Some soils developed on alluvium are classified as Vertisols because of the swelling nature of their clays.

The simplest treatment of alluvial soils is in the FAO-UNESCO (1974) system. It is a mono-categorical system because it is a list of soil units, and soils are not grouped into higher categories at different levels of generalization. Most alluvial soils fall into the Fluvisol category although some, more mature soils, would be classified as Cambisols, Luvisols or Xerosols. There are four divisions of Fluvisols. Eutric Fluvisols have a base saturation greater than 50% between 20 cm and 50 cm from the surface but are not calcareous. Calcaric Fluvisols are calcareous between 20 cm and 50 cm from the surface. Dystric Fluvisols have a base saturation of less than 50% between 20 cm and 50 cm from the surface and Thionic Fluvisols have a sulphuric horizon or sulphidic material, or both, at less than 125 cm from the surface. Thionic Fluvisols are very acid soils, not usually found on floodplains, which generally occur on artificially drained alluvium of mangrove swamps and deltas of tropical rivers (Chapter 8).

Some of the problems of classifying floodplain soils are caused by a lack of agreement on criteria to separate gley soils from gleyed variations of other soils. Gibbs (1980) has suggested that more emphasis should be placed on gleying, but not all alluvial soils are gleyed. Ferrolysized soils also present problems. In the USDA scheme some soils may be Albaqualfs and others Aeric Haplaquepts. In the FAO-UNESCO scheme, if the topsoil and upper subsoil are ferrolysized, such soils would be called Planosols but soils in which only the topsoil is ferrolysized would be classified according to subsoil characteristics. Thus alluvial soils create a number of conceptual problems for the soil scientist. More than most, they require an appreciation of factors concerned with geomorphology, hydrology and climate, as well as pedology (Gerrard, 1987).

CHEMICAL COMPOSITION AND CLAY MINERALS

Alluvial soils in temperate regions often contain calcium carbonate and acid conditions are uncommon. In subtropical areas, with relatively dry climates, the calcium carbonate content of alluvial soils can be high, but in the humid tropics many alluvial soils contain no calcium carbonate and acid conditions prevail. Exceptions to this rule do occur, such as the Ganges–Brahmaputra delta which impinges on deposits rich in calcium carbonate. Because of the more acid conditions of tropical alluvial soils, cation exchange capacities are lower and exchangeable base ratios differ from those of temperate regions. Also, due to the greater destruction of organic matter at higher temperatures, alluvial soils in tropical regions generally have a lower organic matter content. The clay mineralogy of temperate floodplain soils can be comparatively simple. DeMumbrum and Bruce (1960) have shown that montmorillonite was the dominant clay mineral in Mississippi alluvial soils with lesser, but significant, amounts of kaolinite. Clays at depth were basically the same as those at the surface except for some calcite and interstratified montmorillonite-vermiculite, and the large amount of montmorillonite causes some of the soils to shrink considerably on drying.

The alluvial materials of the major river basins may have been transported considerable distances before being deposited. Thus, they are representative of the climatic and geological conditions prevailing in all parts of the basin. This is exemplified in their mineral suites. Some alluvial deposits contain only kaolinite whilst in others the clays are largely montmorillonite. The relationships between climate, weathering, rock type and clay minerals are nicely portrayed in the floodplain soils of Sri Lanka and Egypt. Sri Lanka can be divided into wet and dry zones with a narrow intermediate band between. For the most part, the components in the alluvial soils of Sri Lanka are derived from clay and associated minerals formed by the *in situ* weathering of the largely

Precambrian igneous and metamorphic rocks. In the wetter south-western zone, the situation is complicated by the subsequent weathering of the soils (Herath and Grimshaw, 1971). Gibbsite is common in the wet zone and can be attributed to the tropical weathering conditions. Secondary weathering frequently produces mixed layer minerals. The alluvial clay minerals of the dry zone have been formed by similar primary weathering processes but after transportation and deposition they have not been subjected to the same secondary weathering processes. These soils contain kaolinite and montmorillonite but gibbsite is absent. In the intermediate zone, minor amounts of gibbsite may be present together with montmorillonite.

The complexity of clay mineralogy in some floodplain soils can be seen in the basin of the River Nile (Hamdi, 1959, 1967; Hamdi and Barrada, 1960; Hamdi and Iberg, 1954; Hamdi and Naga, 1950; Hashad and Mady, 1961; Khadr, 1960). The dominant clay minerals appear to be montmorillonite, kaolinite and illite. Until 1961 most published data suggested that the main clay mineral was illite (Hamdi, 1954), whereas after 1961 the significance of montmorillonite has been stressed. The dominance of montmorillonite in the less than 1 micron fraction of Nile delta soils was first noted by El-Gabaly and Khadr (1962). Fishk *et al.* (1976) found that the montmorillonite content varied from 43% to 51% and that of kaolinite from 15% to 20%. Similar mineralogical associations have been reported for the Nile soils of central Sudan and parts of Egypt (Buursink, 1971; El-Attar and Jackson, 1973).

The apparent conflict may have arisen because Hamdi examined deeper samples than most researchers (up to 12 m), and at present-day rates of deposition samples at 12 m in the Nile delta may be up to 12 000 years old. Thus it is difficult to compare surface with deep samples and it may not be sensible to compare the delta areas with alluvium in the middle and upper reaches of the river. It is interesting that illite has been identified in notable amounts in the subsoil of parts of the Blue Nile floodplain (Blokhuis *et al.*, 1970).

A montmorillonite-kaolinite-mica assemblage is accepted by most Egyptian clay mineralogists as being the general composition of the less than 1 micron fraction of lower Nile alluvial soils (e.g. Gewaifel, *et al.*, 1970; Fishk *et al.*, 1976). Illite and chlorite have been identified in some Nile suspended sediments (Nabhan *et al.*, 1969) which may explain traces of these minerals reported in soils by Hanna and Beckmann (1975). It appears that montmorillonite has been inherited from weathering of feldspars in the basalts and basement complex rocks of the source regions of the Nile. It is certainly the dominant clay mineral in the sediments of the Blue Nile and its tributaries in central Sudan (Ruxton and Berry, 1978). Its formation appears to be favoured by relatively dry climates and by topographic and soil situations where water throughflow rates are slow. As noted above an association between dry climates and montmorillonite has been shown in Sri Lanka (Herath, 1962).

There is a negligible contribution of White Nile sediments in the flood-plains of Egypt and most of Sudan. Most of the suspended sediment of the White Nile is deposited in major swamps before the confluence with the Blue Nile (Gewaifel *et al.*, 1981). There are also considerable morpho-logical, chemical and physical variations between the floodplain soils of Egypt and Sudan (Gewaifel and Younis, 1978) reflecting the varied en-vironmental conditions.

The complexities imposed on floodplain soils by climate and parent material variability is well-displayed in the Indian subcontinent. Illite appears to be the dominant clay mineral in Indian floodplain soils but kaolinite, montmorillonite and traces of chlorite have been noted (Sehgal and De Coninck, 1971). Illite and kaolinite are common in soils of the Indus valley (Razzaq and Herbillon, 1979). Swelling chlorites were dominant in subhumid to semi-arid regions, whereas smectite was most common in arid areas. In general, swelling clay minerals were present in greater quantities lower down the soil profiles.

Thus the clay mineralogy of floodplain soils is extremely complicated and there are a number of unresolved problems. Climate exerts a major influence but parent material, distance transported and age are also important. Secondary weathering and the transformation of one clay mineral to another add to the complexity. There is little doubt that to understand the clay mineralogy of floodplain soils requires knowledge of both geomorphology and pedology.

CONCLUSIONS

There is little doubt that soils on alluvial landforms of the world's major rivers, especially those in developing countries, are causing many envi-ronmental problems. Such problems range from flooding and waterlog-ging to salinization and loss of structure – solutions to such problems need to take account of the many relationships that exist between the soils, landforms, materials and river processes. The principles of soil geomorphology are displayed to the full on floodplains.

7

Soils on river terraces and alluvial fans

River terraces and alluvial fans have been grouped together for a number of reasons. They are often composed of similar water-lain materials, possess distinctive and often repetitive landforms, and soil patterns, if not the soils themselves, are often similar. Also in large river valleys, terraces and alluvial fans are often interrelated. Some fans have been formed over river terraces and then later buried by subsequent river deposition. These considerations make it sensible to treat river terraces and alluvial fans together. Inevitably, there will be some overlap with Chapter 10, where desert landforms are considered, however, soils on desert alluvial fans, especially in hot deserts, owe most of their characteristics to the hot dry environment. Alluvial fans in other environments will have followed different genetic pathways. Thus the two treatments reinforce the underpinnings of soil geomorphology.

RIVER TERRACES

Terraces are abandoned surfaces not related to the present stream and are composed of two parts: the scarp and the tread above and behind it. The term 'terrace' is, therefore, usually applied to the entire feature, i.e. both tread and riser. A distinction needs to be made between terraces cut in bedrock and those comprising the former floors of alluvial valleys. The former is the 'strath' or 'rock-cut terrace' in which the riser consists almost entirely of bedrock capped by only a thin veneer of gravel. The other major type, the fill terrace, is composed entirely of sediment laid down during a period of aggradation. Soils developed in the thin deposits of strath terraces in humid regions are likely to be gleyed whereas well-drained soils are more likely on thick gravels. The term 'terrace' is sometimes used to mean the deposit itself when alluvium underlies tread and riser. But it should be referred to as a fill, alluvial fill, or alluvial deposit to differentiate it from the topographic form.

The sequence of events leading to terrace features may include several periods of repeated alluviation and incision. Thus, it is possible to develop any number of terraces and, depending on the magnitude and sequences of deposition and erosion, any number of fills or different stratigraphic units can be deposited. Also, several alluvial fills can be present in the valley even when no terrace exists. In some cases the sequences are simple but very often the terraces and sequences are complicated by hillwash deposits and, in much of Europe and North America, by the deposition of loess. The movement of soil and sediments across the tread and down the steep slope of the riser often makes the topographic features indistinct and the resultant soils very complex. The classic example of this is in the valley of the River Svratka at Cerveny Kopec, Brno, Czechoslovakia. Here a suite of five terraces has been largely obliterated by successive periods of loess deposition, soil formation and hillwash movement. It is estimated that about 900 000 years of loess deposition are represented. In situations such as this, soil relationships are quite complex.

Bull (1990), in providing an excellent summary of river terrace and soil formation, has argued that in studies of soil–landscape interrelations one should be curious as to whether terrace-tread incision is the result of uplift, climatic changes or internal adjustments within the fluvial system. Soil–landscape interrelations are best understood when the genesis of both river terraces and their soil profiles are understood.

Terraces, once thought to be essentially controlled by climatic and tectonic processes, are now seen to be complex-response features (Schumm, 1973, 1977; Schumm and Parker, 1973). Thus for Douglas Creek, north-west Colorado, it is not possible to correlate the terraces purely on the basis of elevation, nor to relate the terrace remnants throughout the valley to specific stimuli such as climatic variations or base-level change (Schumm et al., 1987). The challenge for pedology is to provide a means of distinguishing such terrace types; formation of stream terraces involves changes in the behaviour of the fluvial system. Degradation and aggradation in rivers are separated by the threshold of critical power (Bull, 1979). Aggradation occurs when resisting power exceeds stream power; conversely degradation occurs when stream power exceeds resisting power. In tectonically stable areas aggradation causes temporary departure from the base level of erosion but degradation may return the river to the same level. In such situations soil profiles on terraces tend to be buried and interpretation has to be resolved by following standard stratigraphic principles (Chapter 12).

Where uplift is rapid, flights of river terraces are formed. Thus Quaternary river terraces may differ in altitude by > 50 m where uplift is rapid whereas in tectonically inactive areas terraces may be separated by < 5 m. However, in the former case soil profile development may be similar because terrace formation was closely spaced in time, whereas in the latter case soils on terraces may differ in age by more than 100 000 years even though only slightly separated in altitude.

CHRONOLOGICAL STUDIES

As outlined in Chapter 1, pedologists would like to be able to use terrace treads in detailed chronological studies. But, as Bull (1990) queries, are terrace treads or straths sufficiently synchronous to be regarded as time lines? The paradox is that the pedologist may be the only person able to decipher the enigma. Analysis of morphological and chemical characteristics of soils can be used to evaluate whether differences between pedons, if present, are attributable to different times of formation, altitudinal changes in soil microclimate, or to differences in soil materials. Jackson *et al.* (1982) point out that aggradation surfaces of climatic terraces of large rivers in humid regions are often diachronous because it may take several thousand years for sediment created by climatic change to move downstream – aggradation terraces are only likely to be synchronous for small streams in humid regions. In arid and semi-arid regions there is considerable spatial variation in stream response to climatic oscillations and tectonic activity. This was demonstrated by Weldon (1986) in Cajon Creek, in the Transverse Ranges of southern California, where reaction to Pleistocene–Holocene climatic changes varied enormously. Aggradation began about 17 000 years ago at the mouth of the basin but only reached the headwaters about 6000 years ago. Aggradation had ceased 10 000 years ago for the stretch nearest the mouth but had yet to commence in the headwaters reach. It is very difficult to compare soil profiles on the respective terrace treads because the downstream reach has a subhumid climate and schist-gravel parent material, whereas the upstream reach has a semi-arid climate and sandy parent materials with a substantial amount of atmospheric dust derived from the adjacent Mojave desert. Complex response terraces tend to be regionally diachronous whereas tectonic terraces have the best regional synchroneity.

This discussion has raised two important issues:

1. The conceptual framework for river terraces on which soils form.
2. Terrace features that affect soil genesis.

As Bull (1990) concludes, most terrace soils occur on terraces controlled by Quaternary global climatic shifts, and by internal adjustments within fluvial systems that produce complex-response terraces. Climatic aggradation terraces may provide an ideal flight of surfaces on which to study soil chronosequences. Palaeosols are common in tectonically stable regions where younger aggradation events have been able to spread deposits over treads of older climatic terraces. Complex-response terraces have times of tread formation that must have spatial variability. Thus soils in a spatial sequence will be different unless the time span of formation for the entire terrace-forming event is brief. Climatic terraces of adjacent watersheds should be roughly synchronous. Thus variations in soil properties should be the result of lithologic and climatic factors within the watersheds and not the length of time the soils have been forming. However, correlations of complex-response terraces between fluvial systems are likely to be diachronous and similarities of soil properties may be merely coincidence.

SOILS OF RIVER TERRACES

A good illustration of the relationships between soils and river terraces is that described by Walker (1962b) from Australia. A suite of four terraces exists in many of the steeply-graded drainage basins inland of Nowra, New South Wales. The typical structure of these terraces is shown in Figure 7.1. Terrace K_3 has a wide flat surface 10–12 m above the stream and is composed of layered sands overlain by silts and clays. Terrace K_2 is narrower and occurs 3 m above stream level. It is essentially composed of loam and clay loam although there is a basal gravel layer. Terrace K_1 occurs on the shoulder of number K_2 terrace and is composed of evenly textured loams and clay loams. Terrace K_0 is immediately adjacent to the stream and about 1–2 m above it. The sediments are extremely variable and consist of stratified sands and gravels with occasional bands of silt and clay. The terraces seem to represent simple phases of alternating stream entrenchment and aggradation. Radiocarbon dating suggests dates of 29 000 ± 800 years BP for terrace K_3; 3740 ± 100 years BP for terrace K_2; 390 ± 60 years BP for number K_1 and a modern age for terrace K_0. Walker (1962b) concludes that increased stream aggradation and terrace building were associated with changes to relatively drier climatic conditions.

There has been very little true soil formation on terrace K_0. Depositional laminations are evident and faunal processes have not homogenized the deposits. Terrace K_1 soils have a prairie-type profile with only A and C horizons. The A horizon is a loam to clay loam with strong granular structure and abundant faunal activity. On well-drained areas, faunal activity and organic matter have penetrated to about 1 m, below which depositional structures are evident. There has been very little alteration or translocation of mineral constituents. Soils of terrace K_2 are

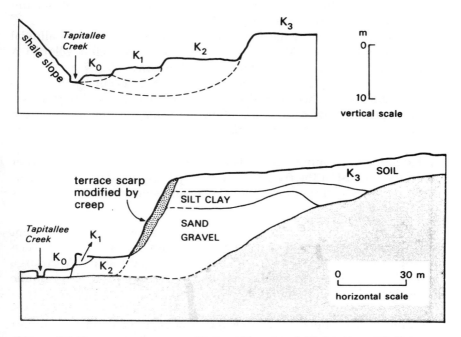

Figure 7.1 Terrace systems near Nowra, New South Wales (from Walker, 1962b).

grey-brown profiles with gradational A/B/C/ horizons. There is a gradual increase in clay content with depth and clay skins coat the voids in the lower B horizon. Soil formation has involved some segregation of mineral constituents. Red and yellow podzolic soils occur on terrace K_3 with a pale, light-textured surface overlying a thick plastic clay B horizon. Segregation of sesquioxides has occurred in the lower A2 and B1 horizons and the B2 horizon colours are reds and yellows with grey mottles at depth. There is a strong segregation of mineral constituents and reduced faunal activity and organic matter penetration.

A general 'model' of soil development on terraces of different ages has been developed by Walker and Coventry (1976) based on a number of studies in eastern New South Wales, Australia. Five general stages are recognized:

1. Stratic stage. This is characterized by soil profiles of low alluvial benches which are frequently flooded. Profiles are essentially unmodified alluvium with prominent sedimentary features.
2. Cumulic stage. These include profiles of well-established floodplains which possess strong organic colouration to depth, numerous faunal features and a lack of sedimentary features. Organic carbon and ni-

trogen tend to be of relatively high values at some depth and calcium ions dominate the exchange complex.

3. Low-contrast solum stage. Such soil profiles are found on low alluvial terraces which are seldom flooded. The soils have well-developed A horizons with gradational changes to B horizons.
4. High-contrast solum stage. Soils of this stage occur on high terraces and are leached with strongly differentiated A and B horizons. They possess low cation saturation, with magnesium and sodium dominant and a clay mineralogy dominated by kaolinite in the < 2 μm fraction.
5. Extended subsolum stage. Soils of this stage are the oldest and occur on the highest of the river terraces. They have pronounced weathering features such as thick alluvial clay and silicates and ferruginous zones.

Soil profiles in the stratic stage are in a state of continuous adjustment to flood deposition and only soil processes with a rapid turnover, such as organic-biotic processes, can approach a steady state. With large inputs of sediment, pedogenesis will be interrupted. In the cumulic stage, incursions of sediment are small and less frequent. It is then possible for profiles to achieve a balance between soil biotic processes and rates of flood deposition. Soils on terraces above flood-level reflect the duration of pedogenesis, although those of the low-contrast solum stage may receive infrequent, small additions of sediment. Walker and Coventry (1976) conclude that a rate-of-growth time scale is appropriate for profiles of the stratic and cumulic stages and a surface-age time scale for profiles of the high-contrast solum and extended subsolum stage. The surface-age time scale is also probably appropriate for soil of the low-contrast solum stage.

One of the most detailed studies of soils on river terraces has been conducted at five sites in the Southern Alps, New Zealand (Tonkin and Basher, 1990). Changes in soil horizonation with increasing soil age and under increasing rainfall are shown in Table 7.1. At Slovens Creek, soils on the youngest terrace (ST1) possess an A/2BC profile with a sandy loam A horizon formed in alluvium. Soils on the older terraces (ST2 to ST4) possess A/AB/Bw/BC/2C profiles. Morphological development from ST2 to ST4 involves increasing B horizon rubification accompanied by clay-content increases from 28% to 36%. ST1 soils have higher pH, lower percentage carbon and a higher total of exchangeable bases than ST2 to ST4 soils, which are all similar in these properties. Soils on the youngest terrace are Orthents and on all the older terraces are Dystrochrepts. At Cave Stream, the sequence of soils is similar to that at Slovens Creek, with A/C soils on the youngest terrace (CT1) and soils with A/AB/Bw/2C profiles on terraces CT2 to CT4. B horizon thickness increases from 20 cm on terrace CT2 to 30 cm on CT4, with a clay content increase from 33% to 39%.

The sequence of soils on the floodplain and seven terraces at Larry River has been described by Ross *et al.* (1977). Soils are formed mainly from Holocene and Late Pleistocene alluvium with some loess on the older terrace surfaces. Profiles change from A/C/2C on the flood-plain (LR1), to A/Bw/2C on the low terraces (LR2 to LR4), to O/A/Eg/Bhs/2C on the intermediate and high terraces (LR5 to LR8). Soil depth increases with age and soil structure becomes increasingly well-developed from the floodplain soils (LR1) to the low terraces (LR2 to LR4). On the older terraces, structure in the Eg horizon is massive. Soils on the floodplain are Fluvents, those on the low terraces are Dystroch-repts, with Haplaquods on the higher terraces.

Table 7.1 Changes in soil-horizon development with time for terrace sequences in some drainage basins in the Southern Alps, New Zealand (from Tonkin and Basher, 1990)

(a) Slovens Creek, eastern region (830 mm a^{-1})

	ST1	ST2	ST3	ST4
Horizon	A	A	A	A
	2BC	AB	AB	AB
		Bw	Bw	Bw
		BC	BC	BC
		2C	2C	2C
Age (ka)	<1	<10	?	18–25

(b) Cave Stream, eastern region (1470 mm a^{-1})

	CT1	CT2	CT3	CT4
Horizon	A	A	A	A
	C	AB	AB	AB
		Bw	Bw	Bw
		2C	2C	2C
Age (ka)	<1	?	?	ca. 8

(c) Larry River, western region (2000 mm a^{-1})

	LR1	LR2–4	LR5–8
Horizon	A	A	O
	C	Bw	A
	2C	2C	Eg
			Bhs
			2C
Age (ka)	<0.5	<14–ca. 20	>22

Table 7.1 *(contd.)*

(d) Wanganui River, western region (6500 mm a^{-1})

	WR1–3	WR4–6	WR7–11
Horizon	A	O	O
	C	Bw	Eg
	2C	C	Bs
		2C	2BC
			2C
Age (ka)	0.05–0.4	0.6–1.6	3–12

(e) Cropp River, western region (10 000 mm a^{-1})

	CRT1–2	CRT3	CRT4–5
Horizon	A	A	A
	C	Agj	Eg
	2C	Bw	Bsm
		C	Bs
		2C	2C
Age (ka)	0.07–0.2	1.3	7–10

Soil has been examined on the floodplain (WR1) and ten terraces (WR2 to WR11) on the Wanganui River. In well-drained sites soil profiles change from A/C/2C on the floodplain and terraces younger than 0.4 ka (WR1 to WR3), to O/Bw/C/2C on terraces estimated to be 0.6–1.6 ka (WR4 to WR6), and O/Eg/Bs/2BC/2C on terraces WR7 to WR11 estimated to range from 3 to 8 ka. The major morphological changes with time are 'the development of surface organic layers, increased soil depth, a change from sandy loam to silt loam mineral surface horizons, increased weathering and disintegration of coarse clasts and the initial development of structure followed by the collapse of structure in the Eg to form a massive horizon' (Tonkin and Basher, 1990, p. 560). Soils younger than 0.4 ka are Udorthents, those between 0.6 and 1.6 ka in age are Dystrochrepts and those estimated to be between 3 ka and 12 ka are Aquods. Soils on terraces along the Cropp River exhibit a similar sequence. Soils on the younger terraces (CR1 and CR2) possess A/C/2C profiles and those on a gently sloping fan (CR3) have an A/Agj/Bw/C/2C profile with an incipient iron pan between the Agi and Bw horizons. Soils on older terraces (CR4 and CR5) have A/Eg/Bsm/Bs/2C profiles. The chronosequence involves a decrease in percentage of gravel and sand and a decrease in pH in surface horizons. Eluvial-illuvial coefficients show an increasing loss of total Ca, P, Fe, Al, Si from surface horizons with time. Soils < 0.2 ka are Udorthents, the soil dated at 1.3 ka is a Dystrochrept and those dated at 7–10 ka are Placaquods.

Table 7.2 Soils on four terraces in the Kennet Valley, England (from Chartres, 1980)

Site	Depth (cm)	Colour (Munsell notation)	Texture	Other characteristics
Furze Hill	0–60	very pale brown (10YR 7/4)	gravel with a sandy silt loam matrix	abrupt, irregular lower boundary
	60–120+	yellowish-red (5YR 4/8)	gravel with a sandy clay matrix	reddened mottles
Hamstead Marshall	0–120 120–170	light yellowish-brown yellowish-red (5YR 5/6)	gravel with a silt loam matrix gravel with a sandy loam matrix	clear, irregular lower boundary reddened mottles
	220–280	strong brown (7.5 YR 5/6)	gravel with some sandy loam matrix material	evidence of fluvial bedding
Thatcham	0–50	light yellowish-brown (10 YR 6/4)	sandy loam with occasional stones (flints)	distinct wavy lower boundary
	50/60–110	reddish-yellow (7.5 YR 6/6)	sandy loam with common stones (flints)	faint reddish mottles
	110+	–	gravel	evidence of bedding at 2 m +
Beenham Grange	0–120/150	light yellowish-brown (10 YR 6/4) to dark brown (7.5 YE 4/4)	silt loam over silty clay loam	merging boundary with underlying gravels
	120/150+	–	gravel	evidence of fluvial bedding

Table 7.3 Phases of soil development on terraces in the Kennet Valley, England (from Chartres, 1980)

Furze Hill	Hamstead Marshall	Thatcham	Beenham Grange	Possible time of formation
aeolian additions cryoturbation: formation of involutions, disruption of existing soils	aeolian additions cryoturbation: formation of involutions and ice wedges, disruption of existing soils	aeolian additions cryoturbation: disruption of existing soil; less marked than at the two upper sites	Terrace deposition	(Late Devensian) Devensian
pedogenesis: lessivage of egg-yellow clays	pedogenesis: lessivage of egg-yellow clays	pedogenesis: lessivage of egg-yellow clays terrace deposition under niveo-fluvial conditions		Ipswichian
				Wolstonian
cryoturbation pedogenesis: lessivage and rubification terrace deposition	cryoturbation pedogenesis: lessivage and rubification terrace deposition			Hoxnian
				Anglian or older

Although considerable attention has been paid to the stratigraphy of terrace sequences in the British Isles few studies of soils on these terraces exist. This may be because several major climatic changes have occurred during the formation of these terrace suites, which makes the relating of pedological differences between terraces to time alone, extremely difficult. One notable exception to this is the study by Chartres (1980) in the Kennet valley of southern England. The soils of four Quaternary terrace levels have been examined in great detail showing major differences between the deposits and soils (Table 7.2). The three upper terraces possess distinct colour and textural discontinuities within the matrix materials. The contact between these two different horizons on the Furze Hill and Hamstead Marshall terraces appears to have been disrupted by periglacial activity.

This study is important for the way it combines detailed analysis of the soils, including micromorphological techniques and heavy mineral assemblages, with geomorphological and geological considerations to produce a synthesis for the evolution of the complete terrace sequence. It is a prime example of the techniques of different disciplines complementing and reinforcing each other. The suggested sequence of phases of soil development is shown in Table 7.3. On the two highest terraces a phase of strong clay illuviation associated with manganiferous staining and rubification of the illuvial features represented the first phase of soil development. This was followed by disturbance of the illuvial features and then a second phase of clay illuviation characterized by the formation of strongly oriented egg-yellow ferriargillans. A second period of disturbance, accompanied by aeolian addition, and a third phase of clay illuviation completes the sequence. A great many features of relevance to Quaternary history may, therefore, be incorporated in polygenetic soil profiles.

The differences between these soils are essentially a function of time although the texture of the sediments has some influence. The terrace soils described by Corless and Ruhe (1955), in western Iowa, represent a different situation as the terraces are mantled by loess and, because of this, formation commenced at approximately the same time. This is equivalent to the pre-incisive sequence of Vreeken (1975). A well-developed loess-mantled terrace occurs along the valley sides of the Nishnabotna River and is probably a remnant of a glacial outwash-alluvial complex. The same series of soils is found on the terraces and on the nearby uplands, therefore it appears that soil-forming factors have had a similar effectiveness on the terraces and the uplands. But differences occur as a result of slight variation in topography. Where runoff is inhibited on the flats and small depressions on the terrace surfaces, the A horizons become thicker and the B horizons become finer textured. This situation is reversed where gentle slopes exist and water can run off rather than accumulate.

Soil formation on river terraces is, therefore, a function of the distinctive landform–material assemblage and the age of the terrace. Where the terrace sequences are relatively straightforward and especially where they can be dated, considerable information is provided about the way soils and soil properties change with time. Conversely, analysis of soil properties is essential to the accurate elucidation of landscape history. Analysis of soils on terrace sequences therefore provides useful information for both pedologists and geomorphologists.

ALLUVIAL FANS

A number of definitions of alluvial fans have been proposed. Thus an alluvial fan has been described as a low cone of gravels, sand and finer sediment that resembles an unfolded oriental fan in outline (Patton *et al.*, 1970), or as a body of stream deposits whose surface approximates a segment of a cone that radiates downslope from the point where the stream leaves the mountain area (Bull, 1968). Perhaps the most comprehensive definition is that of Thornbury. Where a heavily-loaded stream emerges from hills or mountains onto a lowland there is a marked change in gradient with resulting deposition of alluvium, apexing at the point of emergence and spreading out in a fan-like form onto the lowland (Thornbury, 1954).

An interesting study of the relationships between soils and alluvial-fan materials and landforms has been conducted by McCraw (1968) in New Zealand. Alluvial fans are well-developed in New Zealand because a plentiful supply of suitable material exists in the mountain areas and there have been major phases when the material has been transferred to the lowlands and intermontane basins. The area chosen for study was central Otago. A common soil pattern has been developed which is best developed on small, moderately sloping isolated fans. This pattern comprises soils of the fanhead, soils of the middle fan and soils of the toe. Slopes on the fanhead can be as steep as 10° and the surface is irregular and broken by boulders and stream channels. Soils are confined to a small fan-shaped zone extending for less than one-fifth the length of the fan and are shallow. Stony loamy sands overlie coarse porous gravels.

The middle fan zone possesses slopes in the range 2° to 5° and forms a crescent-shaped zone comprising about a quarter of the length of the fan. Soils form a pattern of shallow, sandy loams lying on sinuous gravel ridges running down the radii of the fan, separated by strips of deeper or more finely textured soils. The toe areas slope at less than 2°, occupy at least a half of the fan, and generally possess smooth slopes. Soils are deep silt loams or fine sandy loams developed on silts and fine sands that are compacted and gleyed at depth. Patches of poorly-drained soils are common. This pattern is the result of the spreading out of finer material as the water flow loses its energy when the outer fan areas are reached. The sequence is not unlike that described for crevasse splays in the previous chapter. Some of the waterlogging is a function of the way in which the fans have developed. Fans develop incrementally, and the first small fan would have been buried as successive floods enlarged the feature. Early toe sediments would have been buried by coarser material and new toe sediments deposited further out. Fine sediments would always have been present on the underlying floor. Fine sediments thus become buried and remain as lenses within the sequence of fan deposits; it is these layers that aid the waterlogging process. The porous overlying sands and gravels allow water to percolate from the surface, and this water, when obstructed by lenses of finer material, may move laterally and emerge as seepages and springs.

Sometimes the soil patterns are altered by interactions between the fan systems and the rivers in the main valley. Old levees on the main river may stop the spread of the fan and toe sediments are diverted into old stream channels. Thus a peripheral zone is formed around the toe of the fan where the soil pattern consists of streams of deep soils similar to those of the toe, separated by islands of stony, terrace soils. The levee also interferes with drainage from the fan creating areas of gley soils on the distal parts of the fan toe. In wet periods, swamps and peat bogs may develop.

Analysis of soils on fans throughout the Otago region enabled McCraw (1968) to arrange them into developmental sequences. On fans of similar age, variability in soil types can be related to climate. Brown-grey earths have developed under annual rainfalls of 14 inches; then a sequence from yellow-grey earths to yellow-brown earths and to pod-zolized yellow-brown earths under annual rainfalls of more than 80 inches. In more localized areas soils of fans on different terrace levels form a chronological sequence as follows:

1. Soils on small fans on the river floodplains show little or no profile development. These fans are still very active and soils are either truncated by periodic flooding or are buried by fresh debris.
2. Soils on fans on low terraces are weakly-developed with weak structure in the A horizon and only a slight colour variation to indicate a B horizon.
3. Soils on fans on the intermediate terraces are quite well-developed with compact, olive-brown subsoils and weakly-developed claypans.
4. Soils on fans associated with high terraces possess leached A horizons and strongly-developed subsoil claypans.

McCraw (1968) argues that the soil pattern appears to depend on the amount of entrenchment of the fan stream. With no entrenchment the least-developed soils occur on the fanhead. Where a fanhead trench exists, the least-developed soils are on the middle part of the fan. In situations where the fan stream has incised for nearly the whole of the fan length, the least-developed soils occur at the toe or on a new fan created beyond the toe. This is a classic example of the soil pattern only being understood with reference to the geomorphic processes. Similar conclusions were reached by Butler (1950) in his study of soils on alluvial fans in the Riverine Plain of south-eastern Australia.

The four degrees of soil development on the Otago fans at different levels suggests four main phases of landscape evolution as follows:

1. Fans with little or no soil development are recent and the erosion providing material for the fan was caused by human interference with the vegetation.
2. Fans with weakly-developed soils may correlate with the production of alluvium on the Canterbury Plains which has been dated at about 1000 BC (Cox and Mead, 1963). The erosion may have resulted from destruction of forests after a change of climate (Raeside, 1948).
3. Fans with well-developed soil profiles are associated with intermediate terraces which are related to outwash plains of the last ice age (McKellar, 1960).
4. The large high terraces with well-developed soils are older and their soils have experienced the cumulative effect of a range of climates, possibly including the warm climate of an interstadial.

CONCLUSIONS

River terraces and alluvial fans, with their distinctive landform and material assemblages, provide an excellent natural laboratory within which to examine soils and pedological processes. In this respect, as with floodplain soils, great care has to be taken to differentiate sedimentological from pedological processes. Sequences of terraces are fruitful areas for the pedologist, especially if the terraces can be dated. This allows soils to be placed in a realistic chronology and conclusions reached on the rates of soil formation. The properties of the soils also provide the geomorphologist with invaluable information concerning past climatic regimes and stable/unstable phases in the history of the landscape.

8

Soils of coastal plains and sand dunes

MATERIALS IN COASTAL SYSTEMS

Salt marshes and coastal plains are composed of a variety of materials. These materials have been derived from various sources, transported at various speeds and converted to stable deposits at varying rates. Usually elements of marine, fluviatile and aeolian deposition are all involved. The type of material available for transport and deposition varies according to the texture of the rocks eroded. Thus, the generally soft-rock coastlines of south and east England produce essentially silt deposits whereas the cliffed coastlines of much of north Scotland produce sand deposits. Rivers draining areas of predominantly basic rocks also provide many nutrients which favour the growth of salt marsh vegetation such as *Artemisia maritima* and *Carex divisa*. On other coasts varied geological frameworks and local energy conditions determine the mix of sand and silt sizes.

Much of the material in the coastal systems of the northern hemisphere are reworked glacial deposits which form an offshore store from which supplies are drawn to feed the depositional coastal system. Large tropical rivers produce great quantities of silt, usually originating in chemically-weathered interiors, some of which is deposited on the river floodplains. But most of the sediment is transferred to the offshore coastal systems to be reworked many times before being finally deposited.

There may also be variable inputs of material related to seasonal climatic patterns. Kamps (1962), working in the Eastern Wadden of the Netherlands, reported an increase in clay content from spring to autumn. Increased flocculation due to high salinity and coagulation of particles by increased biological activity in the summer months results in the deposition of large quantities of mud. In autumn, salinity and biological activity decline and greater storm activity leads to a greater influx of silt and sand-sized particles.

The diversity of origins for coastal materials is shown in the evolution of coastal deposits along the Surinam coast, where both fine-textured clays and coarser sands are found (Augustinius and Slager, 1971). These materials have reached the coast by many different pathways. It is thought that the clay originates from the Amazon River and has been transported in suspension along the north coast of South America by a variety of local currents (Reijne, 1961). The sand deposits originate in parts of French Guiana, as well as the Surinam rivers and are moved by offshore and tidal currents. In addition there is a variable input from wind action.

The interaction of the many processes acting in the coastal systems with the variety of materials ensures that a complex series of landforms is created. Many of these landforms are initially unstable media for both soil formation and vegetation growth but eventually, by either upwards growth by accretion, or change in sea level, soil processes can be initiated. The sequence soil formation then follows is determined by the landforms and materials.

COASTAL LANDFORM SYSTEMS

Within the tidal flat complex, vertical tide fluctuations can be used to define geomorphological subenvironments and a number of zones can be identified emphasizing the delicate relationship between landforms, sediments and sedimentary structures (Figure 8.1). In temperate areas, tidal zones are characterized by a complex series of creeks and channels. The offshore marine zone possesses very gentle seaward slopes and is composed of coarse to medium sands. Shell debris may be abundant locally and the clay content increases seaward. The channels and inlets may be up to 600 m wide and 20 m deep near the major inlets, narrowing to 2 m wide and 1 m deep in salt-marshes. Some unusually wide, deep and bluntly-terminated channels are often found behind beach ridges. They were probably connected to earlier inlets but are now filling with finer sediments. Because of the fast currents, channel sediments are coarser than the sediments in adjacent environments.

Bays are generally of two types: large, deep and open bays occur in the widest part of the coastal complex whereas small, shallow bays with less than 50 cm of water at low tide characterize the narrow inland parts. The physical structures of the sediments and biota composition vary with these different settings.

The zone between mean low water and mean high water consists largely of sand, especially in the vicinity of creeks. Some distance away

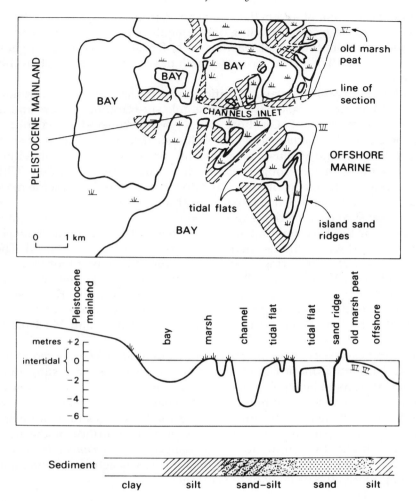

Figure 8.1 The major geomorphic subenvironments of coastal areas (from Harrison, 1975).

from creeks sedimentation of mud and sand takes place and an unvegetated area of flat sand-bars and mudflats forms which is dry at low tide.
At mean high water marked changes occur; there is a drop in current
velocity and the deposition of more clayey sediments occurs. The frequency of flooding also declines allowing the growth of vegetation
which in turn speeds up silting, but, as the frequency of flooding is
further decreased, the sedimentation rate also declines. There is a much
more rapid upgrowth of sediment along the banks of creeks causing an
eventual inversion of relief, and the future soils of these levees will have

a lower clay content and better drainage. Periodic swings of tidal channels ensure large-scale recycling of sediments over comparatively short periods of time. Studies at Caerlaverock, Dumfries, Scotland, suggest that virtually all of a 1500 acre salt-marsh has developed in less than 140 years.

Tidal-flat organisms play a significant role in the process of accretion. Many filamentous algae have considerable silt-trapping powers and in a series of experiments Ginsburg *et al.* (1954) discovered that a culture of the algae *Phormidum* could establish a surface mat of 4 mm of sediment in 24 hours. Molluscs also have an influence on accretion, either by providing surface roughness which enhances sedimentation (Gillham, 1957), or by reconstituting clay into faecal pellets which settle more quickly than unmodified clay particles (Kamps, 1962).

The physical characteristics of the sediments change as the conditions change. In the bare mudflat stage, the material is anaerobic, scarcely consolidated with a limited permeability of the order of 10^{-4} m/d. The withdrawal of water brings about an irreversible dehydration causing the soil to shrink and crack. The sediments also undergo the process of consolidation, either under their own weight or from overlying sediments as, under certain circumstances, the annual accretion can be compensated by the settlement of the sediments (Stearns and MacCreary, 1957).

On tropical coasts, clays are concentrated in mudflats connected to the coast. In South America, these mudflats often consist of a sling mud which appears to be a gel which forms when there is a high concentration of fine particles (Diephuis, 1966). The particles form a structure that settles as a whole and the sling mud gradually changes into an unripe clay by settling of the particles. As soon as the surface of the mud becomes higher than mean tide-level, mangroves (*Avicennia nitida*) start to grow and further aid the depositional process by acting as a filter and slowing down the velocity of the currents.

SOIL FORMATION AND SALT-MARSH DEVELOPMENT

As accretion develops and the sediments remain above the water for longer periods of time, gradual changes in microrelief take place. With levee growth at the edges, the original convex contours of the open mudflat surfaces develop into concave surfaces. Eventually the relief levels off as the higher parts receive less material and the lower areas continue to receive sediment. Pioneer plants, which bind the deposits, are patchy at first and low vegetated hummocks develop, but as sediments accumulate, the plant cover spreads and becomes more varied. Depressions, starting as wide shallow channels, gradually develop into dendritic systems of creeks crossing elevated salt-marsh. The properties of the resultant soils depend partly on the physical and mineralogical constitution of the parent materials, and partly on past and present environmental factors. Parent materials are chiefly transformed by the leaching of soluble salts and native calcium carbonate, by oxidation, reduction and translocation of iron-bearing minerals, and by the development of physical structures and the penetration of organic matter. In areas which are covered for most of the time no oxidation can take place and no soil formation occurs. Areas nearer the shore, especially the higher salt-marshes and creek ridges, are periodically dry and fissures form due to the shrinkage of the clay. But narrow vertical differences in altitudes result in considerable differences in submergence. As Gray and Bunce (1972) have shown for Morecambe Bay, England, the pioneer salt-marsh zone has a mean submergence of about 350 tides, whereas the low-level saltings have a mean submergence of about 200 tides and the high-level saltings about 50.

Unconsolidated materials of loam and clay texture with appreciable reserves of calcium carbonate will favour the development of well-structured soils if they are not permanently wet or saline for too long. Areas permanently above high-water mark suffer fluctuating water-tables. Low water-table levels in summer allow extensive subsurface shrinkage and fissuring which produce strong prismatic and angular blocky peds. The inhibiting effect of poor drainage on structure development is seen in soils developed on creek-bed deposits which possess weak structure in the surface horizons and are structureless after about 30 cm depth. As the material cracks and fissures, stresses are set up which result in the formation of a surface soil considerably denser than might be expected. This stress is the result of unequal water extraction by the roots of pioneer vegetation and of uneven remoistening in the rainy season. To a certain extent this stress is compensated by plastic flow seen in the plasmic fabrics of the subsoil. The mechanism of this process is still largely unknown but, when the soil dries out, the salt content of the surface soil seems to increase considerably and it is thought that as water flows through the fissures and biopores, electrolyte concentration in the soil solution in the vicinity of the pores decreases, peptization takes place and part of the soil mass flows to the subsoil.

When land has been reclaimed and flooding prevented, soil development is largely governed by the seasonally fluctuating water-table levels. Three zones can then be discerned in the soil: the first is below the lowest level of the water-table and is a zone of permanent reduction; above this is a zone which alternates between conditions of aeration and water saturation and therefore between periodic oxidation and reduction; the third zone, nearest the surface, is waterlogged only for short periods and is never appreciably anaerobic. These zones are most clearly differentiated in porous, loamy or sandy materials, as clay soils can have one or more impermeable layers which restrict the percolation of rainwater. Where soil and site conditions are suitable, rapid leaching takes place and salt contents are generally low. Saline soils tend to occur in poorly-drained sites and near the coast where they are subjected to salt spray.

In the early stages of soil formation the bulk of the adsorbed ions will be associated with clay mineral lattices. The cation exchange capacity of the deposits will thus depend on the clay minerals present. Kaolinite has an average exchange capacity of 3–15 mEq/100g, illite 10–40 mEq/100g and montmorillonite 80–150 mEq/100g. Both seawater and sand are deficient in nitrogen but supplies in marsh mud are augmented by fixation of atmospheric nitrogen by blue-green algae. Thus, the nutrient content will vary with distance from the sea and with the vegetation type.

The influence of organisms in the early soil-ripening phase is considerable. In the tidal flats and bays species such as *Ensis* sp., *Solen* sp., *Tegelus* sp., *Callianass* sp., and *Arenicola* sp., occupy discrete burrows in the high current areas of sandy sediments. As the sediment becomes finer, *Crassostrea* sp., *Mercenaria* sp. and burrowing worms become important. Isopods, amphipods, worms and *Mercenaria* sp. are abundant in the very fine clay sediments; they aerate and thoroughly mix the sediments to prepare the material for soil formation. Organisms are also important in the later stages of soil formation. Thus, Green and Askew (1965) have attributed the high fertility of soils at Romney Marsh, Kent, to improvements in drainage caused by the activities of roots, ants and earthworms.

Infiltration rates in the early stages are low, even in sandy marsh soils. This is partly the result of a lack of vertical structures but many other factors are involved. An index illustrating the changes that take place in the soil-ripening process is provided by the n-value of Pons and Zonneveld (1965). The formula is:

$$n = A - 0.2 \, R/L + bH$$

where A is the total water content in grams per 100g of dry soil, L and H are the percentages of clay and organic matter in the dry soil, R is the percentage of mineral particles other than clay ($R = 100 - H - L$)) and b is the ratio of the water-holding capacity of organic matter to that of clay. Values range from 0.3 to 0.4 for normal soils whereas soft, freshly sedimented muds have values between 3.0 and 5.0. Mudflats exposed at low tide have values of about 2.0 and lowest salt-marshes 1.2 to 2.0.

DECALCIFICATION OF MARINE SOILS

The leaching of calcium carbonate from the sediments is one of the more important processes in the early stages of soil formation. But investigations in the Netherlands have shown that the decalcification process is not simple and is made up of a number of components. In some cases older deposits were found to be less decalcified than younger ones. Zuur (1936) first focused attention on this problem and since then many workers have suggested that decalcification during upward growth of sediments can also occur (Edelman, 1950; Bennema, 1953; De Smet, 1954; Van Straaten, 1954; Zonneveld, 1960). Furthermore, these investigations imply that calcium carbonate can also be lost during the transport of sediments. The amount of decalcification will depend on the amount of calcium carbonate present in the original sediments. Verhoeven (1962) demonstrated the existence of a calcium carbonate gradient in young marine mud along the North Sea coast and that the finer fractions contained a fairly constant percentage of calcium carbonate. Bruin and Ten Have (1935) found only small amounts of fine-grained calcium carbonate and Doeglas (1950) could not find any difference in the particle size distribution of beach sand before and after elimination of calcium carbonate. Thus calcium carbonate is subjected during sedimentation to similar sorting processes as the other mineral particles and is not biochemically precipitated as was originally supposed (Edelman, 1950; Bennema, 1953; Zonneveld, 1960).

During the period of upward growth of saltings, calcium carbonate is both supplied and leached out. With the appearance of vegetation the formation and decay of organic matter intensifies as does the production of carbon dioxide. Shrinkage cracks also increase the permeability. Zuur (1936) stresses the influence of permeability on leaching, while Bennema (1953) and Zonneveld (1960) considered the reduction state brought about by the increase in carbon dioxide pressure responsible for rapid decalcification. Alternatively, Van der Sluijs (1970) concluded that variable conditions during silting were the most important causes of present-day variations in calcium carbonate content and depth of decalcification in the marine clay soils of the Netherlands. This means that there should be a close relationship between decalcification, landforms and microtopography.

SOIL–LANDFORM RELATIONSHIPS

Many of the relationships outlined above can be illustrated with reference to specific examples. A good area in which to do this is Romney Marsh in Kent where there have been extensive investigations into the nature of the soils (Green, 1968). The deposits and soil materials are complex but can be classified into a few major categories. The major sand deposit, the Midley Sand, is coarse-textured, clay and silts are scarce and calcium carbonate is generally absent. It is exposed at several localities and forms a series of conspicuous banks which may have been a system of sand banks or dunes which suffered dissection and reworking during burial beneath the clays (Blue Clay) and younger deposits. The clays have a high, 2–50%, silt content and are the soil parent material in the alluvium-filled hollows. Extensive deposits of peat occur at various depths and the presence of stumps indicates that they grew *in situ* during a period of low sea-level (Gilbert, 1933). Wood samples taken from the peat have a radiocarbon date of about 3000 BP (Callow *et al.*, 1964) and samples, probably of birch, from the foreshore were dated at 3040 ± 94 and 3360 ± 92 BP (Green, 1968).

Since the formation of the peat, deposition has produced a variety of sedimentary patterns of contrasting age and complexity. The largest single element is the shingle cuspate foreland formed of successive ridges thrown up during storms (Lewis, 1932; Lewis and Balchin, 1940). Although the shingle is chemically inert, being largely composed of flint pebbles, a succession of plant communities has developed. The development of Romney Marsh during the post-peat period has been very complex with 'deposition of the finer materials under conditions determined by the ever changing configuration of the offshore beaches and the varying protection they provided from the full force of the sea, relative to fluctuations in sea-level, changes in river course . . . and drainage . . .' (Green, 1968, p. 17).

Different landform assemblages or land types can be recognized which are also reflected in the soil associations. These relationships can be illustrated with respect to the decalcified (Old) marshland. One of the most conspicuous features is the creek ridges which are essentially symmetrical in cross-section and form complex dendritic systems. They are thought to represent a system of creeks cut into the thick peat deposits later infilled with clay, silt and sand (Figure 8.2(a)). As the area dried out, falling water-tables allowed the peat between the creeks to shrink leaving the deeper peat under the creeks unaffected. This led to an inversion of relief and former creeks became ridges (Figure 8.2(b)). Local differences in the height and shape of the ridges are due to the different

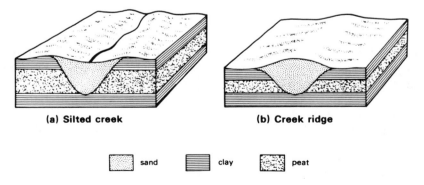

(a) Silted creek   **(b) Creek ridge**

sand   clay   peat

Figure 8.2 Creek ridge development on Romney Marsh, Kent (from Green, 1968).

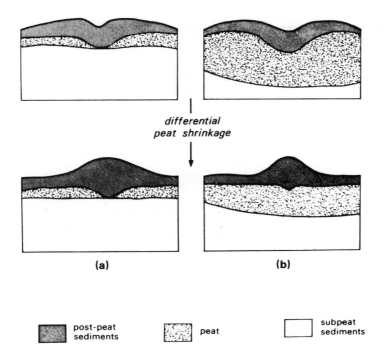

(a)   **(b)**

       post-peat   peat   subpeat
       sediments   sediments

Figure 8.3 Formation of creek ridges of different size and shape (from Green, 1968).

thicknesses of peat; low ridges occur where the peat is thin (Figure 8.3(a)), whereas the narrow ridges developed from deep creeks eroded in thick peat because of greater peat shrinkage (Figure 8.3(b)).

The soil associations of the Old Marshland follow the topographic features very closely and can be differentiated on the basis of texture (Table 8.1). Two sections illustrating these relationships are shown in Figure 8.4. The Appledore Series comprises poorly-drained, non-calcareous clay soils over thick peat which occurs at a depth of 30–60 cm. A typical profile has an Ag horizon, about 15 cm thick, of very dark-grey humose clay with a weak blocky structure. The Bg horizon is of non-calcareous coarsely-prismatic clay. The Dowels Series consists of fine-textured soils over thick peat at 60–100 cm. Silty clay and clay are the dominant textures and, being higher, are better drained than the Appledore Series. The soils of the Snargate Series, which are dominant on the major creek ridges, are moderately well-drained loams or sandy loams to a depth of at least 60 cm below which the proportion of sand often increases to depths of 3 m or more. The Finn Series contains soils

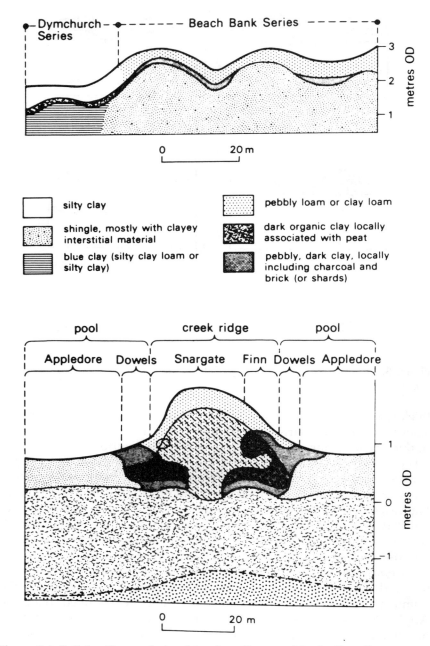

Figure 8.4 Soil–landform relationships from Romney Marsh, Kent (from Green, 1968).

of a finer texture than the Snargate and they are not so well-drained. They replace the soils of the Snargate Series when the ridges are of clay and not sand. Soils of the Dymchurch Series are typically silty clay, though the surface soil to 30 cm may be clay loam or silty-clay loam. The Ivychurch Series is very similar but the soils have coarser-textured horizons between 60 and 100 cm. The structures are coarser or less well-developed compared to horizons at similar depths in the finer-textured soils. They are not plastic unless laminated and tend to occur where sandy lenses are near the surface.

Table 8.1 Texture of soil associations on Romney Marsh (from Green, 1968)

	Silt clay or clay	Clay loam over silty clay or clay	Clay loam	Loam or sandy loam	Loamy sand or sand
texture below 30 in similar or finer than that of B horizon	Dymchurch	Brenzett	Finn	Snargate	Midley
loam to sand between 24 and 42 in		Ivychurch			
thick peat between 24 and 42 in		Dowels			
thick peat above 24 in		Appledore			

These examples demonstrate quite well the relationships between sediments, surface features, drainage and soils. They also show that sea-level changes have occurred quite frequently and have been sufficient to alter the local balance of erosion and deposition. Even with static sea-levels there may be a gradual build-up of material and many coastal complexes are being rapidly regressed. In the Delmarva Peninsula, north-east USA, the rate of regression has been estimated at 10 m a year and if this rate continues the only record of the coastal complex will be in the deeper channels of the inlets (Harrison, 1975). On many coasts the situation is complicated by the fact that a series of transgressions and regressions has occurred in the last 10 000 years. This means that soils are developed on a complex variety of materials in which several peat layers may be sandwiched between sand and silt beds.

SOILS OF COASTAL SAND-DUNE SYSTEMS

Coastal sand-dune systems are characterized by several distinct features. The dunes themselves often show a regular succession from the more active and unstable foredunes at the top of the beach to the older, more stable vegetated dunes inland. The other major element is the system of slacks which exists between the main dune trends. Once a coastal dune has come into being one of two things can happen; either a new dune forms to seaward or the dune grows to its maximum height and then erodes, moving landwards. The first process may add ridge after ridge which eventually stabilize while the eroding dune system may remain unstable for years before becoming permanently fixed. Where prevailing winds are onshore the highest dune will be some way inland whereas the highest dune will be nearest the shore if the prevailing winds are offshore. Rates of dune movement can vary considerably but are often of the order of 3–7 m/yr.

The nature of soils and vegetation has been extensively studied on dunes but less information is available on the slack environment. In the British Isles, the early work of Salisbury (1925) on the sand dunes of Blakeney Point, Norfolk, provided a focus for much later work and in that, and subsequent studies, distinct trends have been identified. Progressive leaching of carbonates occurs with increasing age and coupled with an increase in organic content there is a passage from alkaline to acid conditions. The rate of leaching is rapid at first but then declines as the amount of carbonate rather than the amount of rainfall becomes the limiting factor. The rate of leaching in the early stages depends on the nature of the shell fragments, being slower if the fragments are large. The process will be temporarily halted if material is added to the dune system by wind action and leached layers and organic-rich layers may then become buried.

A detailed study on coastal dune soils by Barratt (1962) in Northumberland, England, amply illustrates these general points. The saline foredunes are building around tufts of sand twitch (*Agropyron junceum*) and humic material is confined to pale-grey patchy stains in the zone of maximum root development a few centimetres below the freshly accumulating sand (Figure 8.5). The grey patches contain rotting leaves which are attacked by white and brown fungal hyphae. The main accumulating dunes above the level of the highest tides are stabilized by marram grass (*Ammophila arenaria*). Dark-grey, root-rich horizons 5–10 cm thick occur at vertical intervals of about 60 cm separated by pale-grey sand. On the consolidated dunes sand is accumulating less rapidly and root-rich horizons appear at only 15 cm intervals. In the dark humic layers, fungal hyphae surround the grains of sand – binding them together in clusters – and play a vital role in stabilizing the root-rich horizons. Vegetation cover is more substantial on the fixed dunes with mosses, lichens and low herbs and a dark grey-brown almost black topsoil up to 2 cm thick containing a mat of fine roots. In places, a vertical sequence of alternating light-grey and dark-grey horizons 1–2

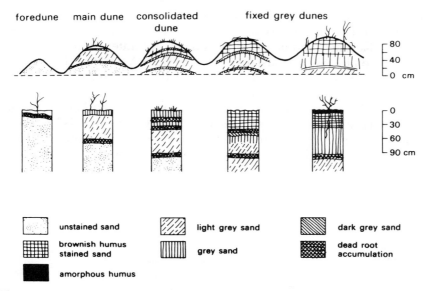

foredune   main dune   consolidated   fixed grey dunes
                        dune

80
40
0 cm

0
30
60
90 cm

	unstained sand		light grey sand		dark grey sand
	brownish humus stained sand		grey sand		dead root accumulation
	amorphous humus				

Figure 8.5 Dune development sequence in Northumberland (from Barratt, 1962).

cm thick extends below the surface to a depth of 8 cm. On the oldest dunes more substantial shrubs occur, the black-brown humic horizon is 8 cm thick and overlies a lighter-brown layer.

The slight differences in soil moisture between the young and old dunes, and between dunes and slacks, control soil formation especially with respect to the oxidation of organic matter. Young slack soils therefore start with a considerable advantage especially as nutrients and other minerals also accumulate in the slack systems. This is ultimately reflected in the soil profiles and pH values. The rate of increase of organic content depends on the initial lime content of the dunes. On lime-deficient dunes, early colonization by *Calluna* is possible with a concomitant increase in the rate of litter accumulation. More acid conditions then prevail, litter breakdown is inhibited and organic matter build-up is promoted.

Old stable dune systems can possess extremely deep soil profiles. Podzols on dune systems along the subtropical east coast of Australia have developed considerable profiles (Thompson, 1983; Thompson and Bowman, 1984). Six dune systems occur. E and underlying Bh horizons increase in thickness with dune age and the oldest soil possesses an E horizon ranging from 11 to 20 m in thickness. The B horizon contains up to 15% clay. It has not been possible to date the soils precisely but the oldest soil has been estimated to be at least 125 000 years old.

Most coastal sand-dune systems can be dated accurately only for several hundred years and information on long-term soil changes is, therefore, limited. But, in some parts of the world extremely old sand-dune systems exist which can be dated reasonably accurately. One such system occurs on the Swan Coastal Plain of Western Australia where three major dune trends can be identified, the oldest being estimated at 200 000 years BP (McArthur and Bettenay, 1960). Total carbonate removal from highly calcareous sand has taken from 100 000 to 200 000 years. Immediately following the total loss of carbonates the pH falls by about two units. There is also a progressive loss of iron with more effective removal taking place in the wetter and less well-drained areas. The complete trends for the Swan Coastal Plain dunes are shown in Figure 8.6.

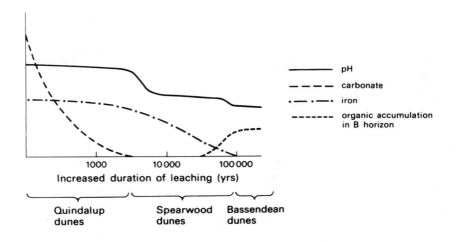

Figure 8.6 Podzolization sequence on ancient dunes in Western Australia (from McArthur and Bettenay, 1960).

CONCLUSIONS

The development of soils on sand dunes and recently exposed marine deposits is a complex process. Both environments are potentially unstable and there is always the interaction between the developing soil profile and the processes of erosion and deposition. Nevertheless, as the brief review in this chapter has shown, certain valid generalizations can be made. The distinctiveness of coastal plain deposits and landforms is frequently matched by the nature of the soils, as vividly seen in areas like Romney Marsh. This allows drainage and reclamation schemes in many parts of the world to be planned accordingly. Where coastal plain sequences can be dated with sufficient accuracy trends in soil formation can be deduced. Also, information obtained from the soil profile allows the identification of periods of geomorphic activity, and the same is true of dune systems although the time-scale involved is often different. In both environments surface stability is accompanied by the progressive leaching of carbonates, a decrease of pH and increase in organic content. Eventually a fully-developed soil profile is achieved. Because of a certain similarity in materials and landforms many of the principles established in this chapter are applicable to areas composed of glacial deposits. These are examined in the next chapter.

9

Soils on glacial and fluvioglacial landforms

Glacial and fluvioglacial deposits exist as a continuous or patchy cover over a large part of the northern hemisphere. In these areas they often form the foundation materials with which civil engineers have to deal, and for these reasons alone the study of soils developed on such deposits is important. The added fact that many of the more recent glacial events can be dated with reasonable accuracy means that information of the rate of soil development can also be deduced. Various names have been used for these deposits such as boulder clay, drift and till. Boulder clay is not a good term because many glacial deposits do not contain boulders and are very rarely composed purely of clay. Drift also tends to include fluvioglacial and glaciolacustrine deposits. Till is by far the most sensible general term to use for glacial deposits and is the one adopted here.

The integration of soil studies with the superficial geology and geomorphology of glacial deposits is beneficial to all. As Scott stresses:

'soils . . . have been intensively studied in Canada, but only a handful of scientists have carried out any quantitative study of till, the parent material whose upper 1 or 2 m are altered to soil. Because of this approach, the effects of regional and local variations in till texture and composition have reached an emphasis clearly secondary to climatic and microtopographic variations in the study of soil-forming processes, the relative emphasis probably being reversed' (Scott, 1976, p. 51).

Space does not permit a detailed examination of all the forms and processes involved in glacial and fluvioglacial deposition. All that is attempted is to highlight the regularities that exist in landforms composed of glacial and fluvioglacial material and to identify the close associations between these and the spatial patterns of the soils themselves and specific aspects of soil formation.

GLACIAL AND FLUVIOGLACIAL DEPOSITS

The characteristics of the deposits vary according to the properties of the environments in which they were transported and deposited. Sugden and John (1976) have argued that there are only three major modes of glacial deposition: material may be released from the ice in the basal zone or at the ice surface by melting, it may be transported on the ice surface and dropped by the melting of the ice beneath it or it may be transported between the basal ice and the solid rock or other underlying glacial drift. Studies of presently active glaciers suggest that the first two processes are the most important. These mechanisms produce tills with widely different particle size distributions, particle shapes, colour, porosity, permeability and compaction. This variability ultimately affects the pattern and types of changes brought about by weathering and soil formation.

Fluvioglacial environments are characterized by fluctuations in discharge over both long and short time periods. Deposits tend to be stratified and the particles are rounded or subrounded in shape. Both deposits and landforms are extremely varied, ranging from the narrow sinuous ridges of eskers to huge expanses of outwash sands or sandar. Any classification system and terminology used in connection with these forms and deposits may have to be amended as our understanding of fluvioglacial environments and deposits associated with existing glaciers is increased (Price, 1973). Fluvioglacial forms and deposits can be differentiated on the basis of depositional environment and whether marginal, or not, to the ice.

Proglacial deposits tend to occur in smoother and more continuous sheets with the sands and gravels being well-sorted and well-stratified. Analysis of deposits in central Scotland by McLellan (1971) bears this out. Outwash sheets show consistently high gravel fractions whereas ice-contact deposits generally contain a wider range of particle sizes and are less well-sorted with large cobbles in association with silt lenses. Strata are liable to be discontinuous and faulted and surface expressions are irregular with ridges, mounds and depressions rapidly alternating.

The type of deposit and landform, and ultimately the soil pattern, will depend on the former relative positions of the ice and the number of glaciations an area has experienced. Deposition in the wastage zone of the glacier is the most complicated because there will be processes of end moraine formation, some dumping of material, some lodgement, melt-out and flowage and ice push. The presence of fluvioglacial and perhaps periglacial processes will add to the complexity. Thus, the sequence of deposits found in formerly glaciated areas can be expected to be quite complex and this should be reflected in the soil patterns.

LANDSCAPES OF GLACIAL AND FLUVIOGLACIAL DEPOSITION

The associations between landforms and materials and their relationships to position within, or in front of, former ice masses enables gross landscape patterns to be identified. Landscape patterns can be examined at different spatial scales, and it is at the intermediate scale where these patterns of depositional landscapes are most readily discernible. Prest (1968), working in North West Territories, Canada, has shown that a zone of drumlins gives way northerly to a zone of elongated drumlinoid forms, then to drumlinized ground moraine and finally to fluted ground moraine. Other workers have shown that at the outer wastage zone, parallel end moraines are conspicuous followed by a zone of ice disintegration features such as kettle holes. On the basis of these and many other consistencies, Sugden and John (1976) have presented a model for

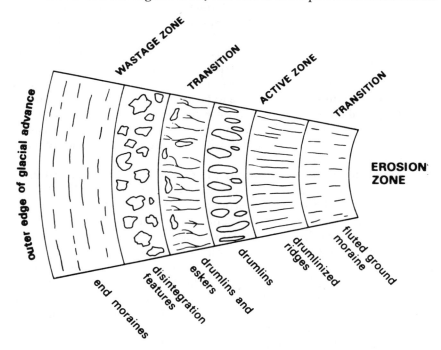

Figure 9.1 Glacial depositional landscape sequence (from Sugden and John, 1976).

the types of depositional landscapes expected to develop beneath part of the periphery of a mid-latitude Pleistocene ice sheet (Figure 9.1). In detail, the situation is more complex but valid generalizations can, nevertheless, be made. Thus, Ruhe (1950) has shown how till sheets of differing ages are characterized by different frequency curves of slope per cent, as the older the till the more integrated are the drainage networks and the steeper are the slope forms. This is reflected in the soils developed on the respective slopes. The distinctiveness of the landform patterns allows predictions to be made of the expected soil patterns. Some of these soil–landscape associations are now examined.

SOILS ON END MORAINES

A conspicuous pattern of swells and swales in a banded arrangement exists in many formerly glaciated areas. Gwynne (1942) has interpreted those of Iowa as being recessional moraines of Late Wisconsin age associated with the seasonal oscillation of a retreating ice front with partial overriding and thickening of material deposited during the season of melting. This landform pattern has a marked influence on moisture distribution and the soil types reflect this. The result is a banded arrangement of the major soil types (Gwynne and Simonson, 1942). Soils of the Clarion Series occupy the swells and possess good natural drainage with three ill-defined horizons. The A horizon (25–30 cm thick) is brownish-black with an intermediate texture whereas the B horizon (10–50 cm) is dark yellowish-brown. Soils in the swales belong to the Webster Series and have restricted drainage. A horizons are thicker (30–50 cm) and are black with a heavy texture while B horizons are dark-grey and thinner (10–25 cm). In this instance recognition of the pattern of moraines has helped the mapping of soils.

Moraines provide ideal environments for studying soil development on hillslopes. First, the approximate age of deposition is known for many moraines, allowing the length of time that pedogenesis and hillslope processes have been operating to be estimated. Second, many glaciers have advanced repeatedly down the same valley creating moraines with similar composition in close proximity. Such moraines have similar parent materials, vegetation and climate but different ages. Thus, on such moraines it is possible to find catenas that are similar in all ways except age, and therefore one may be able to obtain information on the way catenas develop through time (Swanson, 1985). Soil catenas on fine-grained till in humid climates have been extensively studied (e.g. Dalsgaard *et al.*, 1981; Hanna *et al.*, 1975; Pregitzer *et al.*, 1983; Ruhe, 1969; Veneman and Bodine, 1982; Walker *et al.*, 1968 a and b) but Swanson (1985) has noted that catenas on coarse-grained tills in drier climates have received less attention. Exceptions to this have been the studies by

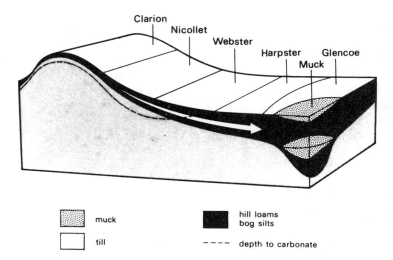

Figure 9.2 The Clarion-Nicollet-Webster soil association in Iowa (reprinted by permission from *Quaternary landscapes in Iowa* by Robert V. Ruhe, 1969, by the Iowa State University Press, Ames, Iowa 50010, USA).

Burke and Birkeland (1983) and Berry (1983, 1984) on moraines in the Sierra Nevada, California and the Salmon River Mountains, Idaho.

On larger moraines, such as the Des Moines lobe in Iowa (Oschwald *et al.*, 1965), these differences in materials and drainage characteristics are sufficient to impart catena-like arrangement to the soils, as suggested in Chapter 3. The relations between soils, such as those belonging to the

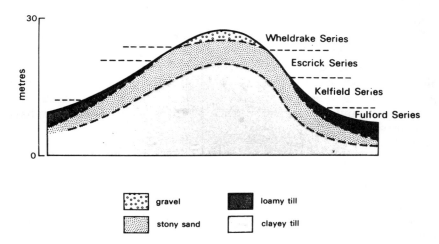

Figure 9.3 Soils on the Vale of York moraine, England (from Matthews, 1971).

Clarion and Webster Series just described, then become clearer. Similarities between areas is so great that Ruhe (1969) has formulated a generalized landscape model with related soil systems (Figure 9.2). The larger moraines in the Vale of York, England, possess similar landform-soil associations (Figure 9.3). Soils of the Wheldrake Series, which are well-drained loamy brown earths, occupy the crests followed downslope in less gravelly situations by brown earths of the Escrick Series. These merge into acid brown earths (Kelfield Series) with, at the base of the slopes, groundwater gleys (Fulford Series). This sequence of soils is common on the larger end moraines of eastern England.

One area where soil development on coarse-grained till in a dry environment has been studied is in the Green River Basin, just west of the Wind River Mountains, Wyoming (Swanson, 1985). Moraines of two ages occur; the older, outer moraine is topographically subdued and contains no undrained depressions, the inner moraine possesses steep slopes, sharp till crests and many undrained depressions. Analogy with nearby moraines at Fremont Lake (Richmond, 1973; Shroba, 1977) suggests that the outer moraines are of Bull Lake age (140 000 years BP) (Pierce *et al.*, 1976). Soils on moraines at Fremont lake have been studied by Mahaney (1978b), Markes (1977) and Shroba (1977). Soils on Pinedale moraines have cambic horizons with slight accumulations of carbonate, clay and silt. Soils on Bull Lake moraines show textural B horizons and stage II carbonate morphology (Chapter 10). In general soils on Bull Lake moraines in the Rocky Mountains contain more stones weathered to grus than do Pinedale moraines (Richmond, 1965).

Similar relationships have been found in the Green River Basin (Swanson, 1985). Soils on the Bull Lake moraine are better developed with A/Bt/C profiles. Also soils on the Bull Lake moraine contain more clay, a higher proportion of pedogenic clay, more weathered stones and more free sesquioxides. Soil maturity is a function of erosion. Erosion on the summits and shoulder sites on the Bull Lake moraine has resulted in thin soils with minimal argillic horizons which are only slightly better developed than analogous soils on the Pinedale moraine. However, on more stable slope positions, especially at concave sites, soils on the Bull Lake moraine are much more strongly developed than analogous soils on the Pinedale moraine.

The soil associations so far described are well-established and soil formation has probably been active for at least the last 10 000 years. In contrast, moraines near presently active glaciers provide excellent natural laboratories within which to chart the course of soil formation. In many cases it is possible to date these moraines extremely accurately and therefore specific time periods can be determined. Studies by Crocker and Major (1955), Crocker and Dickson (1957) and Ugolini

(1968) in Alaska, in the Canadian Rockies (Jacobson and Birks, 1980), in the European Alps (Fitze, 1981), the Elbrus Mountains (Gennadiyev, 1978), Iceland (Boulton and Dent, 1974; Romans *et al.*, 1980) and Scandinavia (Stork, 1963; Alexander, 1970), have done much to increase knowledge of the rate of change of certain soil properties with time. In general pH of the top layers is reduced and calcium carbonate is leached out of the system; nitrogen content decreases and organic carbon increases throughout the age sequences. Vegetation type also influences the speed of these changes with greatest change per unit time occurring under alder. Mellor (1985), in a most intensive study, has examined chronosequences on neoglacial moraine ridges in Jostedalsbreen and Jotunheimen, southern Norway. Four statistically linear models were employed.

$$Y = a + bT$$

$$Y = a + b . \log T$$

$$\log Y = a + bT$$

$$\log Y = a + b . \log T$$

as well as two non-linear functions

$$Y = a - (b/T)$$

$$Y = a / (1 + \exp(C - bT))$$

where Y represents soil properties and T is time. The results for Jostedalsbreen are shown in Figure 9.4. Organic-rich surface horizons display a significant increase with age, although extrapolation of the model suggests a decline in the rate of increase after about 230 years. This is not dissimilar to the results of Jacobson and Birks (1980) who showed that surface organic matter percentages on recent Canadian end moraines increased rapidly for the first 150 years of soil development and then increased more slowly. There is a significant increase in the ratio of the dithionite-extractable iron (Fe_d) in the B horizon to that in the C horizon at Austerdalsbreen. The Fe_d data show a significant tendency towards steady state within 230 years. Also significant increases with age of cation exchange capacity and exchangeable potassium and magnesium (see Figure 9.4(c)) were found in the bleached horizon.

There was no evidence of clay increase with depth, and time-related trends in particle-size data were also absent. However, there is an age-related increase in the thickness of the E horizon implying increase in action of translocatory processes. Also certain elements are moved into it from horizons above (e.g. Figures 9.4(f) and (g)). Organic carbon and pyrophosphate-extractable iron and aluminium showed significant increases with age in this horizon. Not surprisingly pH shows a decrease

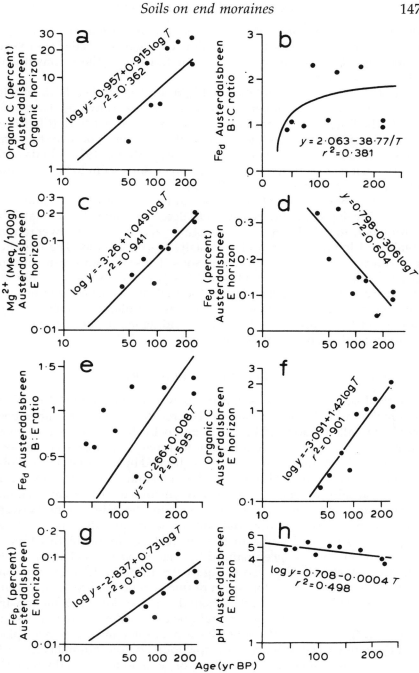

Figure 9.4 Selected soil property chronofunctions showing best-fit equations: Austerdalsbreen, Norway (from Mellor, 1985).

(Figure 9.4(h)). Mellor (1985) argues that the increase in Fe_p and Al_p in the E horizon suggests the formation of organo-metallic complexes whose mobility may be responsible for transfer of these constituents through the E horizon.

There is a significant depletion of Fe_d from the bleached horizon (Figure 9.4(d)) but an enrichment in the B horizon as shown by the ratio of Fe_d in the B horizon to that in the E horizon (Figure 9.4(e)). It has been noted that in the initial stages of podzol formation, iron may be translocated into B horizons in an inorganic form (Moore, 1976; Childs *et al.*, 1983). Such increases in the Austerdalsbreen soils seem to be a function of translocatory rather than weathering processes.

Mellor (1985) has also shown major differences in the soils and their rate of development between Austerdalsbreen, a maritime climatic environment, and Storbreen, a more continental climate with lower rainfall totals and longer duration of seasonally frozen ground. The bleached horizon was absent from Storbreen soils. Thus, the Storbreen soils are Brown soils, as described by Ellis (1979), whereas the soils at Austerdalsbreen are podzols. But, in each case, the analysis of chronofunctions was only possible because of the age-sequence of glacial moraines. More studies of this nature should enable valid generalizations to be made concerning the rate of soil development.

SOILS ON ICE DISINTEGRATION FEATURES

The final stage of till deposition in many parts of the world involved stagnation and ablation of the ice. The resulting landforms have been variously called 'knob and kettle', 'hummocky disintegration moraines', or more generally 'ice disintegration features'. Landforms consist of a chaotic jumble of knolls and mounds separated by irregular depressions. Knolls consist mainly of till, but sand, gravel and crudely stratified drift occur, partially overlain by a layer of reworked drift. Slope stability during and subsequent to deposition and the interactions between slope, moisture distribution and soil formation are all important. The sequence of events leading to the soil features found in this type of landscape has been conceptualized by Acton and Fehrenbacher (1976). Their explanation is depicted in Figure 9.5. In phase 1, before the complete melting of the ice, slopewash of fines would lead to an accumulation of poorly-sorted silts and clays in the depressions. Solifluction and other forms of mass movement could be expected to redistribute some of the till (phase 2). As the ice melts, topographic inversion takes place and these reworked sediments are now found on the summits (phase 3). Continued movement and reworking of sediments by solifluction and mass movement leads to the infilling of these new topographic lows and any soils that were in the process of formation might be removed or

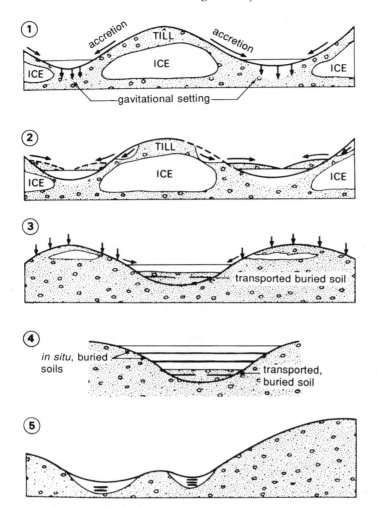

Figure 9.5 Origin of deposits and development of soils in hummocky disintegration moraines (from Acton and Fehrenbacher, 1976).

buried (phase 4). Radiocarbon dating of such soils in Saskatchewan emphasizes that these processes must have taken place very soon after the final ice wastage. Although periods of increased stability and instability allowed a sequence of organic-rich horizons to build up and to be buried later, slope stability eventually prevailed and a modern soil developed (phase 5).

These processes probably apply to the majority of soils developed on hummocky disintegration moraine as similar features are found in many parts of the British Isles. Such a sequence of soils in the Lancashire Plain has been described by Crompton (1966). Brown earths occur on the moderate to steep slopes of the hillocks with profiles which have A or Ap horizons of brown to reddish-brown loamy sand with medium granular structure. The underlying horizons are of strong-brown to yellowish-brown structureless sand which is loose and incoherent. Layers of gravel and stones are common in these horizons. Gleyed brown earths occur on the lower slopes and in some of the better drained hollows whilst groundwater gley soils are found in the enclosed hollows where marked groundwater fluctuations exist. Soils on the hillock crests are relatively thin and represent continued removal of material during formation; soils on the lower positions are often silty and banded with organic matter. These lower soils are developed almost entirely in accreted sediments and are of greater thickness. Erosional and accretionary processes appear to have a greater influence on the resultant soil characteristics than does *in situ* weathering. This was also the overriding conclusion to emerge from the exhaustive work in Iowa by Walker (1966). Successive layers in the centres of the depressions attest to the alternations in stable and unstable phases on the neighbouring slopes.

General trends of soil properties from hillock to depression can thus be identified. Organic matter increases as the thickness of the Ah horizon and the total soil profile increase. Increased leaching and eluviation, as shown by the development of Ae horizons, lower pH values and lower base saturation, occur in the depression soils. Marked Cca horizons are present in upper-slope positions with more diffuse accumulations in the moister lower-slope profiles. Soluble salts are lost from surface horizons and move vertically to underlying horizons and laterally to lower-slope positions, and prismatic structure increases towards the depressions as lime carbonates are removed from B horizons with eventual enhancement of secondary blocky structure in illuvial profiles exhibiting Bt horizons.

These examples emphasize that considerable redistribution of material has taken place in these landform associations. Hollows and depressions are the recipients of silts and clays washed off the surrounding slopes and it is easy to imagine that the soils and superficial materials are dominated by these movement patterns. But not every surface has been subjected to periods of instability and even on more unstable slopes periods of relative stability have occurred. The processes of weathering cannot, therefore, be ignored. Alteration of the surface few metres of glacial and fluvioglacial deposits to produce essentially clayey weathered zones is to be expected, especially on relatively level stable landsurfaces. Thus, there are two main ways in which fine-grained materials can be produced from glacial deposits: one by *in situ* weathering changes, the other by erosion and subsequent deposition. The means of differentiating the two processes and materials have attracted attention for a long time.

GUMBOTIL AND ACCRETION GLEY

Many of the older (pre-Wisconsin) till sheets of North America are covered with grey tenacious clay masses. McGee (1891) was one of the first workers to draw attention to this material on the Kansan tills of Iowa. It was then called gumbo and although variously attributed to fluvioglacial deposition, colluvial processes or loess, McGee thought it was the residue of surface weathering. Kay (1916), in an extremely important paper, proposed the name gumbotil for this material. He defined gumbotil as a grey to dark coloured, thoroughly leached, non-laminated, deoxidosed clay, very sticky and breaking with a starch-like fracture when wet, yet very hard and tenacious when dry, and which was chiefly the result of the weathering of tills.

The nature of the material and the course of weathering has been elaborated by Leighton and MacClintock (1962). Weathering would start promptly after deposition and progress through a series of stages as oxidation, leaching and eluviation took place (Figure 9.6). In the early stages an horizon equivalent to a soil A horizon, develops and clay is eluviated. By stage 2 an oxidized but unleached layer appears. Eventually leaching produces stage 3, carbonates are gradually eliminated and a full soil profile is developed. Once the leaching of carbonates has occurred selective silica weathering becomes important (stage 4) and, with time, the characteristic profile of a gumbotil is thought to be produced (stage 5). The essential characteristic of this weathering sequence is that it occurs under conditions of poor drainage.

This sequence of weathering appears reasonable, given the physical and chemical properties of gumbotil, but some workers still argue that gumbotil has not formed *in situ* but is a fine-wash deposit. Thus, Krusekopf (1948) argues that all of the characteristics of a gumbotil indicate that it is a water deposit, or lacustrine in origin, was deposited as a gumbotil and did not become so subsequent to its deposition. It was weathered and leached when deposited. Frye *et al.* (1960) have complicated the issue by outlining five possible ways in which gumbotil-like material could be developed. They are:

1. Accumulation in shallow pro-glacial lakes of the finest constituents of the debris load carried by the glacier.
2. Slow accretion of fine-textured sediments in swampy or marshy places on a till plain, the materials being derived from weathering of the adjacent slightly higher parts of the till plain surface and moved by sheet wash.

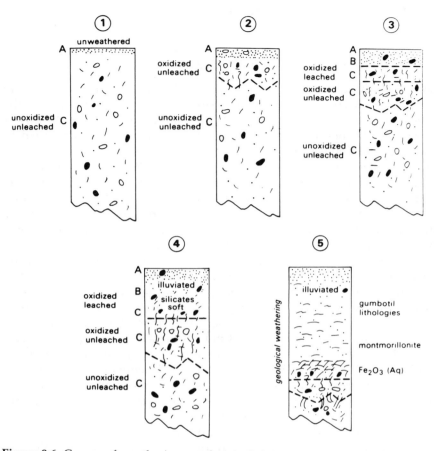

Figure 9.6 Course of weathering on glacial till (after Leighton and MacClintock, 1962).

3. Slow accumulation in marshy environments of fine-textured aeolian sediments intermixed with sheet-wash accretion.
4. Soil profile development in a situation of high water-tables where processes of accretion and *in situ* weathering are both operative; and
5. A process of gleying in a developed soil profile where initial low permeability of the parent material and low topographic relief combine to create poor drainage.

It is readily apparent that a variety of very similar superficial materials but of widely differing origin exists on ancient till sheets. Because of this it is perhaps wise to restrict the term gumbotil to clay masses that have formed *in situ* under conditions of poor drainage. Gleyed material that has accumulated in depressions should be called accretion gley. It is important to be able to distinguish between these materials because of the different information they provide concerning the evolution of the landscape. The presence of gumbotil, formed *in situ*, implies a reasonably stable environment for a considerable length of time. Alternatively, identification of accretion gley demonstrates that there has been redistribution of surface materials.

It should be possible, by using a combination of features, to discover the specific origin of any of these materials. A useful summary of these distinguishing features has been provided by Birkeland (1974). Gumbotil should be massive, without stratification and should be leached of carbonates. Accretion gley would be expected to display prominent layers of contrasting particle size as well as a series of humus-rich layers. The contrast with fresh till underneath should be sharp in the case of accretion gley and gradual with gumbotil. Chemical decomposition and clay mineral alteration will show an orderly arrangement with position in gumbotil profiles, whereas clay minerals in accretion gley deposits will exhibit a disorderly arrangement of different stages of weathering.

The development of the 'classic' gumbotil profile takes a considerable amount of time but even weathering that has been in operation for only a comparatively short period can produce appreciable changes in the upper parts of tills. These changes can be so pronounced that it appears that two tills exist rather than one till with surface weathering. This has been the case in eastern England where Catt and Penny (1966) have shown that many of the features of the Hessle Clay, a supposedly separate and distinct Devensian deposit, that were thought to be diagnostic, could have originated by changes brought about by Postglacial soil development. The Hessle Clay should be regarded as merely a deep weathered mantle on whichever of the two tills occurs at the surface at a particular location, thus invalidating it as a stratigraphic unit.

The changes brought about by weathering, especially in clay mineralogy, also cast doubt over the use of specific clay minerals as either stratigraphic indicators or as till source indicators. Thus, Quigley and Ogunbadejo (1976) have shown that the high smectite contents in the surface tills of parts of North America do not reflect a different till sheet nor a different source area, but *in situ* oxidation and leaching that has removed all the original chlorites and carbonates respectively. Detailed understanding of the weathering of tills is, therefore, essential to a correct interpretation of materials, stratigraphy and Quaternary history.

CONCLUSIONS

Pedological techniques and investigations considerably enhance the geomorphological and geological examination of till provenance and stratigraphic history. Conversely, geomorphological concepts are invaluable in assessing the significance of particular soil features such as stone lines and humus layers, and demonstrate that a considerable redistribution of material, and many landscape changes, have occurred since the last glacial period. Soils and landforms exist in tandem and must be treated as such.

10

Soils and desert landforms

INTRODUCTION

Deserts are complex places. Thus

> the formulation of simple and comprehensive generalisations
> about the nature of desert geomorphology . . . will be difficult, and
> in the present state of knowledge is impossible. The major ob-
> stacles to such attempts lie in the enormous climatic, edaphic,
> biological and hydrological diversity of deserts, the impress of past
> climatic changes, and the roles of endogenetic processes, rock types
> and geological structures (Cooke and Warren, 1973, pp.6–7).

These statements are still basically true, but research over the last twenty
years has begun to unravel some of these complexities. This has involved
multidisciplinary studies linking geomorphology, geology, biology and,
not least, pedology. The prime example of such a study is the Desert
Soil-Geomorphology Project of the US Soil Conservation Service conducted between 1957 and 1972 in the Basin and Range area of southern
New Mexico. The results of this study are examined in greater detail later
in this chapter, but the main message conveyed by this study is that the
complexities of desert landscape history can be understood by examining
the nature of desert soils and their relationships with desert landforms.

Deserts have been classified in many ways. The definition adopted
here, in line with Cooke *et al.* (1982), is that produced by Meigs (1953).
It includes both hot and cold deserts and covers no less than a third of
the world's land surface. Such a definition includes the sand seas of the
Sahara, the mountainous coasts of Chile, the limestone plateaux of South
Australia, the high-altitude deserts of Inner Asia, and Death Valley in
California well below sea-level. A great variety of landforms occur in
such diverse areas but it is possible to establish a number of generaliza-
tions.

DESERT LANDFORMS

A number of comparatively simple desert landform assemblages can be identified. Mabbutt (1969, 1971) has shown that 96% of desert areas in Australia can be classified according to five main types of surface conditions, namely mountain and piedmont desert, shield desert, clay plains, stony and sand desert. However, a different classification scheme for Australian drylands has been proposed by Perrin and Mitchell (1969, 1971). A similar analysis of surface types in four major desert areas has been conducted by Clements *et al.* (1957). But, as Cooke and Warren (1973) point out, the categories used in all these classifications are not necessarily mutually exclusive nor collectively comprehensive.

Tectonic activity has been a major factor in determining gross assemblages of desert landforms. In areas of recent tectonic activity a landscape of dissected mountains, steep mountain fronts, alluvial fans, bajadas and playas occurs (Figure 10.1(a)). This is basin and range topography and occurs typically in the south-western United States but also in parts of Iran, Pakistan, the high-altitude deserts of Asia, the Oman Peninsula and parts of the Andes in South America. In areas where tectonic activity has been less intense or absent, desert landscapes are dominated by plains with bedrock pediments and extensive alluvial and aeolian plains (Figure 10.1(b)). The pediments either form coalescing plains or extend away from isolated hill masses or inselbergs. Many desert areas of Africa, India and Australia possess such landscapes. Horizontally bedded or gently dipping rock strata tend to create a landscape of extensive structural plains, escarpments, mesas, buttes and pediments (Figure 10.1(c)). Rivers tend to flow in deep canyons. The Colorado Plateau, south-west USA is the classic example of such a landscape but other areas can be found in North Africa, the Middle East and Australia. Many studies have established close relationships between sequences of landforms, and the nature of soil and superficial materials (Figure 10.2). This is in reality a geomorphic catena. Processes in desert landscapes also interact to produce meaningful and comparatively simple patterns (Figure 10.3). It is in the context of such landform and surface material characteristics that desert soils have evolved.

DESERT SOIL FORMATION

A number of factors influence the nature and development of desert soils. Extremes of climate have a major influence (Table 10.1). Average precipitation is probably not a relevant parameter to use when considering desert soils as desert rainfall tends to be highly variable in space and time. The interannual variation can reach 100% and many areas receive a whole average years rainfall in one storm (Table 10.2). Relationships

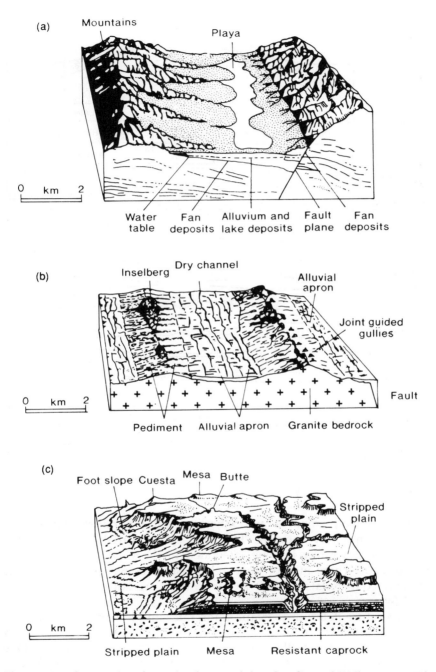

Figure 10.1 Contrasting desert landscapes (after Goudie and Wilkinson, 1978).

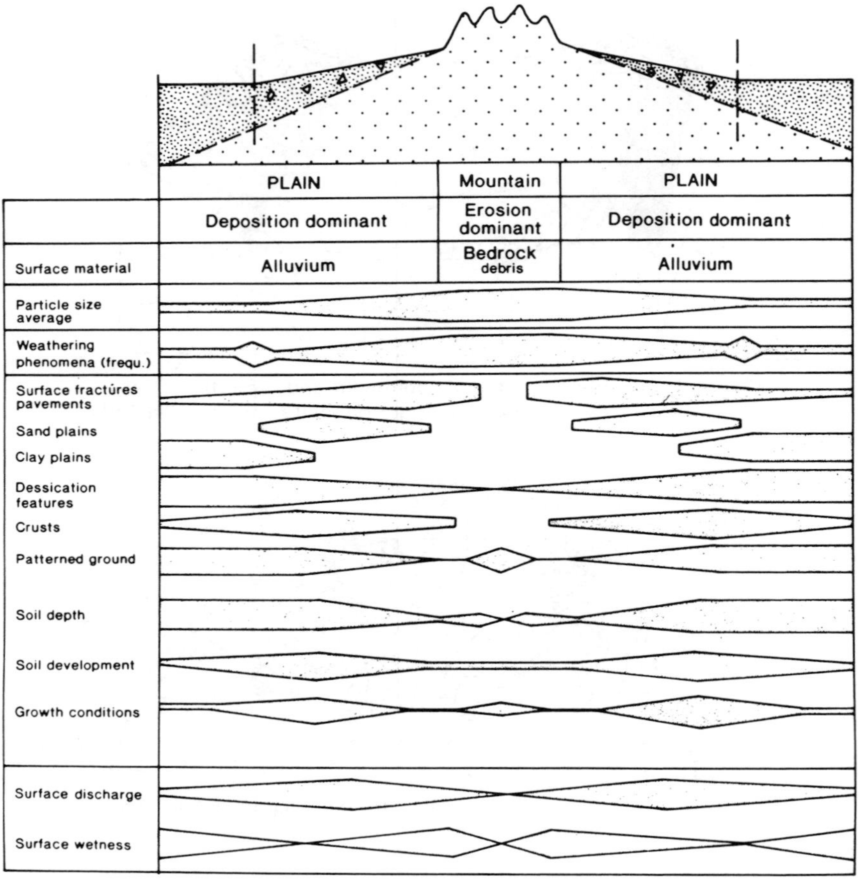

	PLAIN	Mountain	PLAIN
	Deposition dominant	Erosion dominant	Deposition dominant
Surface material	Alluvium	Bedrock debris	Alluvium
Particle size average			
Weathering phenomena (frequ.)			
Surface fractures pavements			
Sand plains			
Clay plains			
Dessication features			
Crusts			
Patterned ground			
Soil depth			
Soil development			
Growth conditions			
Surface discharge			
Surface wetness			

Figure 10.2 Relationships between soil and landform elements along desert slopes (after Cooke and Warren, 1973).

between rainfall amounts and intensity, soil infiltration and surface runoff become very complex. The great spatial variability means that soil landscapes rarely become integrated in the way they do in more humid areas. There may be good relationships between soil, landforms and surface materials but soil–landform units are not necessarily linked spatially.

Temperature determines evaporation rates and how much of the precipitation remains on or in the soil. A number of indices have been developed linking rainfall parameters and intensity of solar radiation. One of the most commonly used it that devised by Budyko for the preparation of aridity maps for the 1977 United Nations Conference on desertification.

(a)

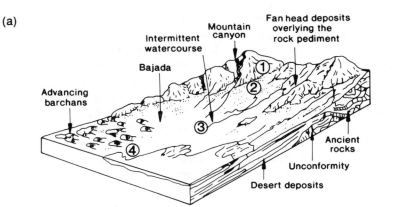

ZONE	1	2	3	4
PRINCIPAL ENGINEERING SOIL TYPES		Rock fans	Silty stony desert and sandy stony desert. Some evaporites	Sand dune loess and evaporites
SLOPE ANGLE OF DESERT SURFACE		2-12°	½ – 2°	0– ½°
PRINCIPAL TRANSPORTING AGENT OF THE ENVIRONMENT		Gravity and as 3	Intermittent stream flow and sheet floods	Wind and evaporation
GEOTECHNICAL FEATURES		Good for foundation and fill	Generally very good foundation and fill material	Erratic behaviour to load bearing. Migrating dunes Metastable loess Saline Absence of coarse material

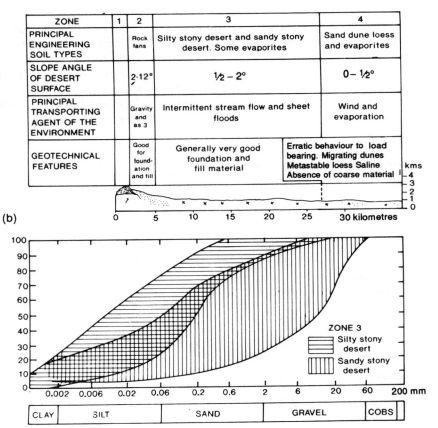

(b)

Figure 10.3 Soil–landform units in deserts (from Fookes and Knill, 1969).

Soils and desert landforms

Budyko's dryness ratio is defined as the number of times the mean net radiation at the earth's surface in a year can evaporate the mean precipitation. On this basis an area will be classified as a desert if the dryness ratio was more than 10, arid or desert margin if between 7 and 10 and semi-arid if between 2 and 7 (Table 10.2). The nature and amount of vegetation reflects the rainfall characteristics to produce distinctive features of deserts (Table 10.2). Desert vegetation has low biomass and low productivity. Biomass production in central Asia can be as low as 40–140 kg/ha which is at least ten times less than for temperate forests. In many desert areas biomass production is zero, whereas in other areas pockets of quite high productivity occur. Great spatial variability is the overwhelming characteristic (Greig-Smith and Chadwick, 1965; Whittaker *et al.*, 1968). A relatively high proportion of biomass production occurs just below the surface in the rooting zone (Cloudsley-Thompson and Chadwick, 1965). This will aid soil formation in the top few centimetres. However, few desert soils have organic matter contents greater than 2% and most possess less than 1%. The nature of desert vegetation also influences soil formation. Algae often form a resistant crust, affecting infiltration rates but also fix nitrogen and help to produce low carbon/ nitrogen ratios characteristic of many desert soils (e.g. Bolyshev, 1964; Fuller *et al.*, 1960; Mayland and McIntosh, 1963; Shields *et al.*, 1957). Potassium and calcium are also accumulated more by desert than humid vegetation.

Table 10.1 Climate data for drylands (from Middleton, 1991)

Station	Average annual precipitation (mm)	Maximum precipitation in 24 hrs (mm)	Average daily temperature (°C)	Average relative humidity (%)
Bilma (Niger)	22	49	26.6	27
Jidda (Saudi Arabia)	76	140	27.7	56
Kashgar (China)	86	25	12.2	58
Antofagasta (Chile)	8	38	16.2	74
Multan (Pakistan)	170	155	26.0	48
Phoenix (USA)	184	78	21.4	43
Alice Springs (Australia)	252	147	20.6	37
London (UK)	593	60	10.7	73

Table 10.2 Characteristic features of drylands (from Middleton, 1991)

Type	Dryness ratio*	Precipitation (mm)	Precipitation interannual variation*	Vegetation	Land use
Hyper-arid	>10	Very low (<25)	100% or more	Very little or none permanent. Some after rain or dew	Oasis culture. Nomadism
Arid	7–10	From 80–150 to 200–350. Low humidity	50–100%	Sparse. Found in water channels	Pastoralism. No farming unless irrigated
Semi-arid	2–7	From 300–400 to 700–800 with summer rains. From 200–250 to 450–500 with winter rains	25–50%	Savannah or steppe grass. Some thorny shrubs	Rainfed cultivation and sedentary livestock
Sub-humid	<2	Abundant with usually more than six humid months	<25%	Grasses and woodlands	Rainfed cultivation and industrial crops

*Note: these terms are defined in the text

Climate and vegetation, in combination, create distinctive characteristics of desert soils, such as low clay content, low exchange capacities (Scott, 1962), and often high-base saturation (Whittaker *et al.*, 1968). Aridity also creates conditions for surface additions by wind deposition and atmospheric fallout (Eriksson, 1958). Much of this addition is oceanic salt. In Israel Yaalon (1963) has shown that salt accumulates in the drier soils more than the wetter soils and salt fall-off curves have been plotted (Yaalon and Lomas, 1970). Much quantitative work has been conducted in Australia with estimations that 168.96 kg/ha/yr of NaCl and 14.28 kg/ha/yr of gypsum were being added to the soil of the York Peninsula, South Australia. The addition of oceanic salt may also be important several hundred miles from the Australian coast (Halls-worth and Waring, 1964), although calcium generally increases as a percentage and sodium declines away from the coast (Hutton, 1968; Hutton and Leslie, 1958). In central Asia, sulphates and carbonates are the most important land-derived salts (Tsyganenko, 1968). Dust incorporation into soils can also be significant (e.g. Smith *et al.*, 1970), with the addition of clay particles being especially significant (Dan and Yaalon, 1968; Smith *et al.*, 1970). The clay deposit in Australia known as parna is thought to be of aeolian origin (Butler, 1956; Butler and Hutton, 1956).

Horizons of carbonate accumulation are characteristic of soils on the semi-arid margins of deserts. An idealized profile of a well-drained desert soil is shown in Figure 10.4. The terminology follows that of Gile *et al.* (1965), and relies on the recognition of a 'K-fabric'. The K-fabric is fine-grained carbonate which occurs as a relatively continuous medium. The carbonate content at which the continuity of the K-fabric is broken occurs anywhere between 15 and 40%. Carbonate accumulation occurs in other layers but carbonate content is usually below 20% and there is less than 50% K-fabric. The K–horizon is divided into three zones; an upper transitional K1 horizon, the K2 horizon which is the carbonate horizon proper, and a lower horizon transitional to the C horizon, the K3 horizon. Indurated, massive K2 horizons are designated K2m and banded or laminar horizons above the K2m are known as K21m. K2 horizons of New Mexico have been shown to have carbonate contents between 23 and 60% and K21m horizons carbonate contents of 75% (Gile *et al.*, 1966). K21m horizons in Tunisia (Coque, 1962) and Texas (Reeves and Suggs, 1964) were found to have carbonate contents of 75% and 91% respectively.

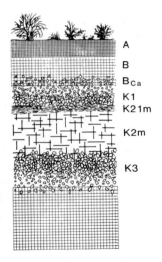

Figure 10.4 Idealized desert soil profile.

Most pedologists now believe that K-horizons have been formed by the downward eluviation and deposition of carbonates. Detailed studies have shown that horizons generally occur at nearly constant depths in the soils, even when the landscape surface undulates (e.g. Gile *et al.*, 1966; Ruhe, 1967; Stuart *et al.*, 1961). The correspondence between relief of the K horizon and surface relief is closer than it would be if the carbonate had been deposited at or near a water-table.

Datable sequences of geomorphic surfaces have allowed the development of K horizons to be established. Such sequences, for gravelly and non-gravelly materials, in New Mexico, are shown in Figure 10.5. Gravelly material contains more than 50% gravel. Other sequences have been described in South Africa (Goudie, 1971), Morocco (Ruellan, 1968) and Texas (Hawker, 1927). With continued carbonate accumulation most or all pores become filled by carbonate, primary grains have been forced apart, bulk density has increased and infiltration rate decreased. This results in the plugged horizon which develops in the last part of stage III. The depths of the plugged horizon correspond to the depths of frequent wettings by unsaturated flow. Once the plugged horizon has formed the laminar horizon can form on top of it. The laminar horizon has much more carbonate than the plugged horizon and rather than the carbonate being a filling between skeletal grains, it occupies almost the entire horizon. The laminar horizon is a new soil horizon and the overlying horizons will have been displaced upward.

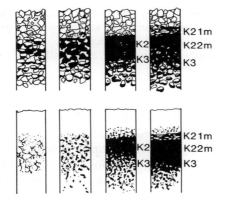

Figure 10.5 Sequence of carbonate accumulation in (top) gravelly soils and (bottom) non-gravelly soils (after Gile *et al.*, 1966).

The numerous laminae indicate that accretion is an episodic process reflecting many wetting and drying cycles. Radiocarbon carbonate dates indicate that carbonate of the plugged horizon is oldest, that carbonate of the lower, hard laminae is more recent and that carbonate of the upper, soft laminae is the youngest (Gile *et al.*, 1981). Some dates also indicate that many laminar horizons must have formed primarily during wetter times in the Pleistocene.

The four morphological stages of carbonate accumulation have been recently expanded to six stages by Machette (1985). Stages V and VI are recognized by the degree of repeated brecciation and recementation. The C horizon is finally subdivided into a slightly oxidized horizon and a visually unaltered horizon. The characteristics of all six stages are shown in Table 10.3.

Table 10.3 Stages of carbonate accumulation in desert soils (from Machette 1985)

Stage	Paleosols developed in gravel	Paleosols developed in sand, silt or clay
i	Thin discontinuous coatings of carbonate on underside of clasts	Dispersed powdery and filamentous carbonate.
ii	Continuous coating all around and in some cases between clasts: additional discontinuous carbonate outside main horizon	Few to common carbonate nodules and veinlets with powdery and filamentous carbonate in places between nodules
iii	Carbonate forming a continuous layer enveloping clasts: less pervasive carbonate outside main horizon	Carbonate forming a continuous layer formed by coalescing nodules: isolated nodules and powdery carbonate outside main horizon
iv	Upper part of solid carbonate layer with a weakly developed platy or lamellar structure capping less pervasively calcareous parts of the profile	
v	Platy or lamellar cap to the carbonate layer strongly expressed: in places brecciated and with pisolites of carbonate	
vi	Brecciation and recementation as well as pisoliths common in association with the lamellar upper layer	

A major problem concerns the source of the carbonate as many K horizons occur in relatively acid parent materials (Brown, 1956; Crocker, 1946; Gile *et al.*, 1966). Ruhe (1967) has also stressed that if the carbonates were released solely by weathering, appreciable amounts of clay would also have been created. Downslope movements of carbonate can occur (Gigout, 1960) and some has originated from percolating streams or floodwaters (Glennie, 1970; Motts, 1958) but in totally insufficient amounts. Observation and measurements have now shown that sufficient carbonate-rich dust is added to A horizons by atmospheric fall-out to account for K horizons (e.g. Bettany and Hingston, 1964; Gile *et al.*, 1966; Jessup, 1960 a, b; Ruhe, 1967).

Geomorphic surfaces, dated by radiocarbon techniques, have enabled the path of soil development to be established. Some of the most comprehensive work has been reported by Gile and Hawley (1968) from southern New Mexico. Soils were examined on well-drained sites where annual rainfall events varied from 200 to 300 mm. The sequence of soil development was extended back to the mid-Pleistocene by extrapolation to surfaces in the nearby Rio Grande Valley. The general sequence of soil development is shown in Table 10.4.

A great amount of information on soil evolution in deserts has been gleaned from examining relationships with geomorphic surfaces. Cooke and Warren (1973) have suggested that Walther Penck's Aufbereitung concept can be applied to desert soils. If surface erosion is minimal a weathered mantle will develop and deepen until a stage is reached where further deepening is slow. Deepening will then continue at about the same rate as slow surface erosion. This will allow a soil profile to develop. They argue that the frequently found result of soil development in deserts is, first to weaken the soil surface, in this way supplying large amounts of material to the aeolian and fluvial erosional systems, and then to provide a hardened carapace which preserves ancient desert surfaces. Such an interplay of ideas is exemplified in the classic study of the Basin and Range area of southern new Mexico, which is now examined.

Table 10.4 Soil ages and characteristics in New Mexico (after Gile and Hawley, 1968)

Soil age in years (soils developed in materials of ages	Features of soil development
100 years (?)	Thin grey A horizon, vesicular in places; slight organic matter accumulation; original sedimentary bedding still present.
100 (?) to 1000 years	Very slight carbonate accumulation; most of the original bedding has been destroyed, and only thicker beds of fine earth remain.
1100 to 2100 years	No distinct signs of carbonate accumulation in soils with little gravel, but in gravelly soils a weak but distinct horizon of carbonate accumulation in the form chiefly of pebble coatings.
2200 to 4600 years	In non-gravel soils a weak carbonate horizon has developed in the form of coatings on structure units, and a distinct structure has developed.
Late-Pleistocene	Orientated clay coatings on grains in a B horizon with clay enrichment, and a K horizon. A Km horizon has developed with a single laminar horizon on its upper edge.
Mid-Pleistocene	The same with a distinct Km horizon with two or more laminar horizons (K21m) on top.

BASIN AND RANGE, SOUTHERN NEW MEXICO

The Desert Soil-Geomorphology Project of the US Soil Conservation Service in Dona Ana County, southern New Mexico, is a classic example of an integrated pedogeomorphological study. It encompasses many of the themes which form the basis of this book. Geomorphic processes, soils and landforms are related to provide a coherent analysis of landscape evolution during the Quaternary Period. The area contains alluvial landforms, terraces and geomorphic surfaces that have been analyzed, in a general way, in other chapters. Buried soils also occur and provide vital clues to landscape history.

The Desert Project covers a 400 sq. mile area studied by a team of soil scientists and geologists from 1957 to 1972. The project region is a 'typical' basin and range desert landscape with the landforms being the result of a complex set of tectonic and gradational processes. Tectonic activity has been the main factor to influence landscape development on the regional scale for the last several million years. But the primary factor controlling geomorphic processes in individual basins and valley segments have been short-term fluctuations in regional climate. Weathering, soil formation and fluvial activity have been affected and mountain glaciers and pluvial lakes have waxed and waned. Two major landform groupings occur (Figure 10.6):

1. Intermontane basins (bolsons) with central plains of aggradation, including lake and alluvial basin floors.

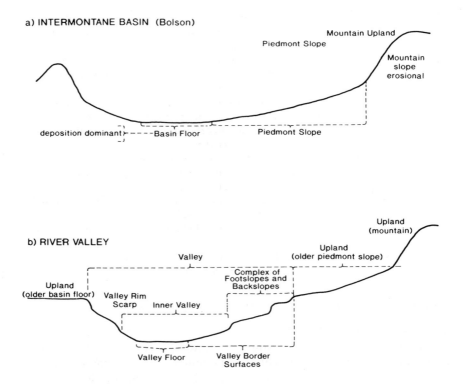

Figure 10.6 Basic landscape types in the Basin and Range area of southern New Mexico (after Gile *et al.*, 1981).

2. River valleys entrenched into older basin fill and rocks of mountain uplands.

Low relief forms superimposed on this fabric are provided by desert pavements, coppice dunes and scarplet erosion surfaces.

Intermontane basins contain three main components: erosional slopes cut in the rocks of the surrounding mountains, flanking detrital slopes of aggradation and a central-lake plain or playa. In the context of this study, two major subdivisions have been employed, namely piedmont slopes and basin floors (Figure 10.6). Piedmont slopes are gently- to moderately-sloping, graded surfaces extending from mountain fronts to nearly level, central-basin areas. A distinct break in slope, the piedmont angle, usually occurs at the piedmont slope–mountain front junction. Basin floors are essentially level alluvial and lacustrine plains occupying central-basin areas. Piedmont slopes are substantial landform elements and range in width from slightly less than 1 mile to 13 miles (1–20 km). Basin floors range from narrow drainage ways several hundreds of feet (about 100 m) wide to broad plains 20–30 miles (30–50 km) across.

Individual landforms on piedmont slopes include rock pediments, alluvial fans and coalescent fan complexes. In internally-drained basins aggradational piedmont landforms are dominant and original depositional surfaces are well-preserved. Erosional landforms, such as pediments and dissected alluvial slopes occur immediately adjacent to mountain fronts. Where mountains are adjacent to large through-drainage systems, erosional forms occupy large areas. A pediment is an erosion surface of low relief that cuts across the rock formations of the mountain front and is usually fringed by an alluvial apron or an erosional surface developed on old alluvium. Piedmont erosional forms are governed by concentrated runoff from large mountain drainage basins. An important role is also played by weathering and erosional processes on upland interfluves and footslopes adjacent to the mountain front (Moss, 1977). Some of the pediments in the present landscape show evidence of having been buried and later exhumed (Oberlander, 1974). They appear to have been buried in late Tertiary to early Quaternary times with partial exhumation in middle to late Quaternary time. Some pediment forms, therefore, may relate to processes acting in environments considerably different from those of the past million years.

Alluvial fans are conspicuous features in the Desert Project region and vary from large mountain front features more than 40 sq. miles (100 sq. km) in area and hundreds of feet thick to small arroyo and gully-mouth fans occupying areas of a fraction of a square mile and comprising deposits less than 10 feet (3 m) thick. Where alluvial fans coalesce laterally downslope a fan piedmont or bajada is formed. In the project region this is the most extensive piedmont surface.

Interfan valleys are locally prominent features at the upper edge of the fan piedmont between the strongly convex, proximal parts of individual fans emanating from major canyons and opposite intercanyon mountain fronts. These triangular-shaped lowlands are occupied by smaller landforms comprising basin-fill and bedrock erosion surfaces and valley-fill units including fans and terraces. Three types of entrenched piedmont channels exist in the internally-drained basins. The first is the fanhead trench (Chapter 7) which is simply the extension of the mountain stream cut through the proximal parts of the fan. The second type occurs in the interfan lows and starts in smaller mountain front drainage basins. The third type consists of swales and incised channel systems that commence on piedmont slopes. Discontinuous gullies with fans and headcuts are extensively developed (Heede, 1974; Patton and Schumm, 1975). Soil development on such features will be determined partly by material composition. Piedmont slopes possess a great variety of deposits but generally coarsen towards the mountains. Most of the variability in textures and sedimentary structures is due to the wide range in sediment-to-water ratios in piedmont-sediment transport systems (Bull, 1962, 1972).

Individual landforms of intermontane basin floors include ephemeral lake plains, relic forms of ancient lakes, both recent and ancient features of alluvial origin, and a variety of aeolian landforms. In the project region the alluvial plain is the dominant basin-floor landform. Alluvial flats are also locally important. These are broad, nearly level surfaces that carry storm runoff from piedmont slopes towards playas. Axial drainage ways are a third type of alluvial landform.

A variation of the basic Basin and Range structure is created by the large, deeply-incised valleys of major through-going river systems such as the Rio Grande and Colorado. A typical Basin and Range river valley (Figure 10.6(b)) possesses a nearly level valley floor and complex sideslopes designated valley borders. The morphology of valley cross-sections ranges from narrow, sideslope-dominated segments to broad, floor-dominated segments. Channel forms are highly variable ranging from deep, narrow, meandering or 'suspended load channels' to shallow, wide, straight 'bedload channels' (Schumm, 1968, 1972). Mixed load river channel forms, such as braids also occur. Along major river valleys a stepped sequence of surfaces occurs between the floodplain and valley rims. These are relic surfaces graded to former river levels and have been designated valley-border surfaces.

SOIL CHARACTERISTICS

In the Soil Taxonomy system (Soil Survey Staff, 1975) Aridosols are dominant in the arid zones between the mountains and Mollisols occur in places along the mountain fronts. Examination of soil moisture in the Jornada Experimental Range has shown that most soils in the arid areas have aridic (tonic) moisture regimes (Herbel and Gile, 1973). Soil-moisture regime is defined in terms of a soil-moisture control section. If the particle-size class is fine-loamy, coarse-silty, fine-silty or clayey the control section is between 10 and 30 cm from the surface; if the particle size is coarse-loamy the crucial zone is between 20 and 60 cm; and between 30 and 90 cm if the particle size class is sandy. Soil is defined as dry when the moisture tension is 15 bars or more. In the aridic moisture regime a soil in most years is dry in all parts more than half the time that the soil temperature at a depth of 50 cm is above 5°C and never moist in some or all parts for as long as 90 consecutive days when the soil temperature at a depth of 50 cm is above 8°C

A horizons of desert and mountain soils in the project region differ considerably. In mountain soils the horizons are dark and contain significant amounts of organic carbon. The organic-carbon content and thickness of the A horizons qualify them as mollic epipedons. This contrasts with the desert soils at low elevations that are thin, light in colour and contain little or no organic carbon. Thus they qualify as ochric epipedons.

Most soils have formed in slope deposits or alluvium. Soils formed on weathered bedrock only occur on the mountain fronts. The intermittent and variable nature of sediment deposition ensures that relations between soils and parent materials are complicated. Many soils contain material eroded from older soils upslope, and some of these earlier soils may have formed under different climates. Aeolian input adds a further complication, and increasingly it is being recognized that atmospheric inputs to soils should not be underestimated (Pye, 1987). As mentioned earlier prominent horizons of carbonate accumulation occur in many desert soils formed in parent materials very low in calcium. Some of the calcium could have originated in atmospheric dust. In the eastern Mojave Desert, California, Wells and McFadden (1987) have shown how the accumulation of aeolian silt and clay reduces soil permeability and lowers soil infiltration capacity. Such aeolian deposition may increase hillslope instability by promoting debris flow activity which provides the matrix needed for slurry flows. Also the presence of salts will accelerate mechanical weathering (McFadden and Tinsley, 1985) such as has been observed on basaltic lava flows in the eastern Mojave Desert (Wells *et al.*, 1985).

In the Desert Project region dust traps were used to assess the magnitude and nature of the atmospheric input (Gile and Grossman, 1979). Clay content of the dust varied from 20% to 40%, organic carbon ranged from 2.5% to 6.6% and carbonate content from 1.3% to 5.7%. The general conclusion of this study was that the movement of labile calcium in dust over a long period of time was an important agent for the dissemination of carbonate over the landscape. This was reflected in the increasing amount of carbonate with increasing age of soils in noncalcareous parent materials. Ionic calcium in the precipitation plus labile calcium in dry dust would have been sufficient to form 2 kg/sq.m of carbonate per thousand years, assuming that all the calcium enters the soil and is deposited as carbonate (Gile *et al.*, 1981).

Soil moisture is affected by landscape position, microrelief and soil morphology. Moisture infiltration is increased by cracks in soil horizons that are connected to holes and depressions at the surface. Soils without such surface features wet uniformly and the depth of wetting is less. Thus, depth of wetting exhibits great spatial variability. In the project region the most favourable moisture conditions were found in soils and landscapes with level or nearly level areas of a stable landsurface, a superficial horizon, about 5 to 10 cm thick, with sand or loamy-sand texture and a slightly finer-textured horizon just below the surface horizon to trap the infiltrated water.

Carbonate horizons are characteristic of soils in the project region. The carbonate horizons are thought to be of pedogenic origin because:

1. the horizons approximately parallel the soil surface;
2. in the arid areas their upper boundaries are near the soil surface and are or have been within reach of wetting;
3. their morphology is distinctive and differs markedly from that of adjacent horizons;
4. the horizons form in morphogenetic sequences related to time;
5. the depth to the horizon of carbonate accumulation increases towards the mountains in soils of stable sites.

As noted earlier, most of the calcium must have come from atmospheric additions. Calcium enters the soil when it is wetted, calcium bicarbonate in solution moves down though the soil during wetting cycles and calcium carbonate is deposited on drying. Radiocarbon dating has established that many of the laminar horizons formed during wetter times in the Pleistocene. Reddish-brown and red horizons of silicate-clay accumulation are also common. Analysis shows that they contain illuvial clay and are Bt horizons. In the more acid areas the Bt horizons range from 15 cm to 1 m in thickness with their upper boundaries commonly 5–10 cm below the surface. Some of the clay has been produced by weathering of igneous rocks in the mountains and some is included in atmospheric dust. Early work on desert soils suggested that clay accumulation in arid soils was due to inplace weathering (Nikiforoff, 1937; Brown and Drosdoff, 1940). Later, more detailed work has shown that much of the clay is illuvial in nature in spite of an absence of clay skins (Gile and Grossman, 1968; Smith and Buol, 1968; Nettleton *et al.*, 1969,

1975). The absence of clay skins seems to result from unfavourable conditions for their formation and preservation.

SOIL–LANDFORM RELATIONSHIPS

A number of geomorphic surfaces of varying age and nature have been identified in the Desert Project region. The valley border surfaces are essentially depositional landforms with fans and terraces predominant (Table 10.5). The piedmont slope areas possess a greater variety of surfaces with fans, terraces, arroyos and dunes (Table 10.5). Materials and landforms are more consistent in basin-floor areas, being dominated by alluvium but with the occasional area of lacustrine sediments – remnants of Pleistocene lakes (Table 10.5). Surfaces in all zones range in age from recent to Middle Pleistocene. A distinction can be made between materials and surfaces related to the regional system of through drainage which resulted in widespread basin aggradation in Pliocene to Middle Pleistocene times and the subsequent entrenchment of the Rio Grande and the development of closed-basin landscapes (Ruhe, 1962). In the early phases depositional environments included internally drained and open hydrologic systems. Bolson environments with playaplains and adjacent aggrading slopes developed for a period of about 20 million years beginning in early Miocene times (Seager, 1975). Differential earth movements in Late Miocene to Pliocene times led to the coalescence of individual basins and the isolation, by partial burial, of many mountain areas. By Late Pliocene the ancestral upper Rio Grande had developed along a chain of structural depressions extending from southern Colorado to northern Mexico. From Late Pliocene to Middle Pleistocene, rapid fluvial deposition occurred leading to a variety of basin and piedmont-slope facies (Seager *et al.*, 1971; Seager and Hawley, 1973; Hawley *et al.*, 1969). The main geomorphic surfaces developed on these materials at this time were the La Mesa, Dona Ana and Jornada I. Ruhe (1964, 1967) has defined the La Mesa surface as a widespread basin-floor surface, underlain by a thin veneer of loamy sediments on Middle Pleistocene rounded gravels and sand of mixed composition. The Dona Ana surface comprises remnants of pediments and alluvial fans. The Jornada I surface (Gile and Hawley, 1968) is the youngest surface that predates the entrenchment of the major rivers. This surface comprises remnants of the original depositional surface, piedmont erosional surfaces cut in bedrock and basin fill and alluvial-flat surfaces associated with thin basin-floor fills.

The surfaces and materials that postdate the onset of Rio Grande incision are extremely variable. The Jornada II materials and surface are the result of episodes of piedmont-slope gradation. The bulk of the unit is a sheet-like fan-piedmont deposit that forms a thin but nearly continuous mantle on many piedmont slopes. Isaack's ranch unit appears to be the result of one or two episodes of piedmont-slope gradation. Organ deposits are Middle to Late Holocene in age, and radiocarbon dates indicate that initial deposition started before 6400 years BP. A variety of depositional landforms are involved. During the Historical period a number of piedmont slopes have been dissected by arroyos and gullies, and the introduction of livestock may have been a contributory factor in this dissection (Buffington and Herbel, 1965; York and Dock-Peddie, 1969). Some of this erosion has created the Whitebottom surface. The Lake Tank surface is an ephemeral playa-like plain. A number of surfaces are associated with the Rio Grande Valley. The Tartugas surface comprises erosion surfaces on bedrock and basin fill adjacent to the mountains. The Leasburg surface consists of one or two minor erosional benches cut into Picaho fan and river deposits before deposition of Fillmore alluvium. The Fillmore surface is the culmination of prehistoric activity of arroyos graded to local base-levels near the present valley floor (Gile *et al.*, 1981).

Table 10.5 Soils, landforms and materials associated with geomorphic surfaces in the Basin and Range region, southern New Mexico (after Gile *et al.* 1981)

Geomorphic surface	Physiographic location and soil age (yrs BP or epoch)	Landform and material
The valley border		
Arroyo channels	Historical (since 1850)	Arroyo; alluvium
(Coppice dunes)	Historical	Dune; eolian sediments
Fillmore	100 to 7000	Fan, ridge remnant, terrace; alluvium; colluvium
Leasburg	Earliest Holocene-Latest Pleistocene (8000–15000)	Fan, ridge remnant, terrace; alluvium
Fort Selden	(Fillmore and Leasburg–undifferentiated)	Fan, ridge remnant, terrace; alluvium; colluvium
Picacho	Late Pleistocene (25000–75000)	Fan, ridge remnant, terrace; alluvium; colluvium
Tortugas	Late to middle Pleistocene (150000–250000)	Fan, ridge remnant, terrace; alluvium; colluvium
Jornada I	Late middle Pleistocene (250000–400000)	Fan-piedmont, ridge; alluvium remnant
La Mesa	Early to middle Pleistocene (400000–1500000)	Basin floor; alluvium
The Piedmont slope		
Arroyo channels	Historical	Arroyo; alluvium
(Coppice dunes)	Historical	Dune; eolian sediments

Geomorphic surface	Physiographic location and soil age (yrs BP or epoch)	Landform and material
Whitebottom	Historical	Small fan, drainageway; alluvium
Organ	100 to 7 000	Fan, fan-piedmont, terrace, drainageway, valley fill, ridge; alluvium; colluvium
III	100 (?) to 1100	Gully fill at Gardner Spring
II	1100 to 2100	Valley fill and terrace at Gardner Spring
I	2200 to 7000	Valley fill at Gardner Spring, extensive elsewhere in landforms noted for Organ
Isaacks' Ranch	Earliest Holocene–latest Pleistocene (8000–15 000)	Fan, drainageway, ridge; alluvium
Jornada II	Late Pleistocene (25 000–75 000?)	Fan, fan-piedmont, terrace, ridge; alluvium; colluvium
Jornada I	Late middle Pleistocene (250 000–400 000)	Fan, fan-piedmont, ridge remnant, terrace
Jornada	(Jornada I or Jornada II, undifferentiated)	As for Jornada I and II
Doña Ana	early to middle Pleistocene (greater than 400 000)	Fan, ridge remnant; alluvium
Basin floor north of US-70		
Lake Tank	Present to Late Pleistocene	Playa; alluvium, lacustrine sediments
Petts Tank	Late Pleistocene (25 000–75 000)	Basin floor; alluvium (lacustrine in part?)
Jornada I	Late middle Pleistocene (250 000–400 000)	Basin floor; alluvium
La Mesa	Middle Pleistocene (greater than 400 000)	Basin floor; alluvium
Jornada I-La Mesa	(Jornada I or La Mesa, undifferentiated)	Basin floor; alluvium

Soils are closely related to the nature of the materials and geomorphic surfaces (Table 10.6). The most complex soil patterns are found in the dissected landscapes, where soils have been truncated to varying degrees. The patterns are complex because some soils have been truncated by erosion and because dissection has caused abrupt changes in particle size. Soils can be examined in terms of gross landscape assemblages.

Table 10.6 Relationships between soils and geomorphic surfaces, southern New Mexico (after Gile et al., 1981)

Geomorphic surface	Age, years BP or epoch	Physiographic position	Soil order or great group	
			Low-carbonate parent materials	High-carbonate parent materials
Arroyo channel (Coppice dunes)	Historical	Channel of arroyo	Entisols	Entisols
	Historical	Valley border; basin floor; lower and middle piedmont slopes	Torripsamments	none
Fillmore	100 to 7000	Valley border	Entisols camborthids Haplargids	Entisols
Organ	100 to 7000	Upper piedmont slopes	Haplustolls, Argiustolls	Haplustolls
		Lower and middle piedmont slopes	Entisols, Camborthids, Haplargids	Entisols, Calciorthids
Leasburg	Early Holocene–latest Pleistocene	Valley border	Entisols, Camborthids, Haplargids, Calciorthids	Calciorthids
Isaack's Ranch	Early Holocene–latest Pleistocene	Lower and middle piedmont slopes	Haplargids	Calciorthids
Lake Tank	Holocene and late Pleistocene	Playa	Torrerts, Haplargids	none
Picacho	Late Pleistocene	Valley border	Haplargids, Paleargids, Calciorthids, Paleorthids	Calciorthids, Paleorthids
Jornada II	Late Pleistocene	Lower and middle piedmont slopes	Haplargids, Paleargids, Calciorthids, Paleorthids	Calciorthids, Paleorthids
Petts Tank	Late Pleistocene	Basin floor	none	Calciorthids, Paleorthids
Tortugas	Late to middle Pleistocene	Valley border	Calciorthids, Paleorthids	Calciorthids, Paleorthids
Jornada		Upper piedmont slopes	Argiustolls, Paleustolls	Calciustolls, Paleorthids
Jornada I	Late middle Pleistocene	Basin floor	Haplargids, Calciorthids, Paleorthids	none
		Lower and middle piedmont slopes	Haplargids, Paleargids, Paleorthids	none
La Mesa	Middle Pleistocene	Basin floor	Haplargids	none
Doña Ana	Middle Pleistocene	Upper piedmont slopes	Paleargids, Calciustolls, Paleorthids, Paleustolls	none

Valley border soils and surfaces

The geomorphic surfaces of the valley border are characterized by their stepped sequence and variable degree of soil truncation. The Coppice dunes are Historical and show little soil development. Soils of the Fillmore surface demonstrate the initial phase of soil horizon development and must have formed under a climate similar to the present one. All soils show a stage I horizon of carbonate accumulation. Relationships between geomorphic and carbonate horizon development are shown in Table 10.7. Upper horizons of soils on stable sites on low-carbonate parent materials are usually non-calcareous. Water infiltration has been sufficient to remove most of the carbonates in the dustfall. All soils formed in high-carbonate parent materials are strongly calcareous.

Table 10.7 Stages of carbonate accumulation on surfaces of different ages, southern New Mexico (from Gile *et al.*, 1981)

Stage		Youngest geomorphic surface on which stage of horizon occurs and age – years BP or epoch	
Nongravelly soils	Gravelly soils		
Arid (valley border)			
I	I	Fillmore	100 to 7000 yrs
II	II, III	Leasburg	early Holocene–latest Pleistocene 8000 to 15 000 yrs
III	III, IV	Picacho	late Pleistocene 25 000 to 75 000 yrs
III	IV	Jornada I	late to middle Pleistocene 250 000 to 400 000 yrs
IV		La Mesa	middle Pleistocene > 400 000 yrs
Arid (fan-piedmont)			
I	I	Organ	100 to 7000 yrs
II	II	Isaacks' Ranch	early Holocene–latest Pleistocene 8000 to 15 000 yrs
III	III, IV	Jornada II	late Pleistocene 25 000 to 75 000 yrs
III	IV	Jornada I	late middle Pleistocene 250 000 to 400 000 yrs
Semi-arid (canyons in the Organ Mts.)			
	0 or I	Organ	100 to 7000 yrs
	I	Jornada II	late Pleistocene 25 000 to 75 000 yrs
	I	Jornada I	late to middle Pleistocene 250 000 to 400 000 yrs
	III, IV	Doña Ana	middle Pleistocene > 400 000yrs

Soils of the Leasburg surface are intermediate between the weakly developed soils of the Fillmore surface and the better-developed soils of the Picacho surface. Illuvial clay is clearly evident in soils developed in low-carbonate parent materials on stable sites. Low-gravel soils of the Leasburg surface show stage II of carbonate accumulation whilst high-gravel soils show stage II or early stage III. Soils of the Picacho surface must have developed, in part, during the last glacial period, and are distinctive for many reasons. They possess a prominent argillic horizon and have substantial amounts of carbonate and clay accumulations. Some soils in gravelly materials show a transition from stage III to stage IV of carbonate accumulation.

The Tartugas surface is not extensively developed and soils developed on surface remnants have been truncated during various erosive phases. This truncation makes it difficult to make any general statements concerning soil development. Some truncation of soils occurred on parts of the Jornada I ridges but, as truncated areas grade into more stable surfaces, pedogenic generalizations can be made. Soils of Jornada I ridges are better developed than those of Picacho and Jornada II surfaces, for instance the Bt horizons have more prominent maxima of silicate clay. In gravel parent materials, carbonate horizons possess multiple laminar horizons, whereas single or discontinuous laminar horizons are common in soils of Jornada II or Picacho age. But, in low-gravel soils on Jornada I surfaces stage IV carbonate horizons have not formed.

The soils on the La Mesa surface are the oldest in the valley border zone. In low-gravel parent materials stage IV carbonate accumulation has occurred. Soils in high-gravel materials show a transition from stage III to stage IV. Soils on the upper La Mesa surfaces appear to be older and have developed complex stage IV carbonate horizons with multi-cyclic laminar zones.

Piedmont slopes, soils and surfaces

Terraces at several levels are conspicuous and the age of both terraces and soils increases with increasing elevation of the terrace treads. Terraces related to Organ, Jornada I and II and Dona Ana surfaces have been identified. Where terraces merge downslope complex soil patterns exist. Alluvium associated with Fillmore and Organ surfaces is of similar age, however soils of the Organ surface are better developed than those on the Fillmore surface. Soils on the Organ surface in the arid zone possess stage I carbonate horizons. Horizons of carbonate accumulation gradually deepen towards the mountains as precipitation increases. Upper horizons also become darker in colour and Aridisols in the arid zone grade to Mollisols in the semi-arid zone. Soils of the Isaack's Ranch surface have stage II carbonate accumulations. Soils in low-carbonate

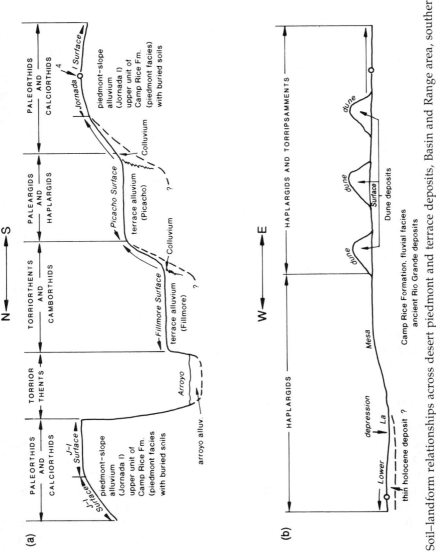

Figure 10.7 Soil–landform relationships across desert piedmont and terrace deposits, Basin and Range area, southern New Mexico (after Gile *et al.*, 1981).

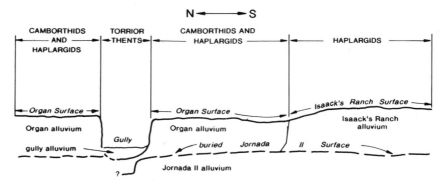

Figure 10.8 An example of buried soils and geomorphic surfaces, Basin and Range area, southern New Mexico (after Gile *et al.*, 1981).

parent materials also have argillic horizons. Soils are intermediate in morphology between those of Holocene age and Late Pleistocene age and are similar to soils of the Leasburg surface.

Soils of the Jornada II surface occur on alluvial fans and coalescent fan-piedmont slopes. In the arid zone, soils are similar to those of the Picacho surface with similar horizons of clay and carbonate accumulation. Bt horizons in soils of the Jornada II surfaces are redder and thicker than those of Isaack's Ranch and Organ surfaces. Carbonate horizons are also stronger being at stage III in low-gravel materials in the arid zone and stage IV in high-gravel materials. Soils of the Dona Ana surface in mountain canyons possess laminar horizons and stage IV carbonate accumulation. On one stable remnant soils have the reddest and most clayey Bt horizon in the Desert Project region.

Basin floor soils and surfaces

Soils of the Lake Tank surface vary widely in age as sediments range from recent to Late Pleistocene in age; soils in the central part of the playa have also been affected by shrinking and swelling. Soils of the Petts Tank surface are of Late Pleistocene age, are strongly calcareous and lack argillic horizons because of calcareous parent materials. The soils possess structural B horizons and stage III carbonate horizons. Soils of Jornada I are older than, and are buried by, sediments and soils of the Petts Tank and Jornada II surfaces, but are younger than soils of the Middle Pleistocene La Mesa surface.

The mosaic of soils and surfaces can be seen in a variety of cross-sections. In some areas terraces, piedmont slopes and arroyos occur in a relatively simple sequence (Figure 10.7(a)). In such situations it is comparatively easy to establish sequences of soil and landform develop-

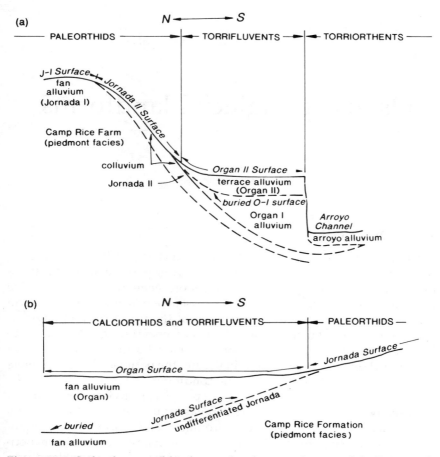

Figure 10.9 Soils, slopes and landscape development in part of the Basin and Range area, southern New Mexico (after Gile *et al.*, 1981).

ment. Simple sequences occur on the lower La Mesa surface (Figure 10.7(b)). As noted earlier, fluvial desert landforms often get buried by later deposits. This is especially true of alluvial areas (Figure 10.8) and produces a sequence of buried soils (Chap. 12). Numerous examples occur in the project region (Figure 10.9(a) and (b)) and provide a means of establishing long-term evolution. Sequences on alluvial fans have also been studied in the area by Gile and Hawley (1966). The whole project has demonstrated that analysis of soils with landforms and pedology with geomorphology is essential to a full appreciation of landscape complexity.

11

Soils and periglacial landforms

INTRODUCTION

The term periglacial was coined by Lozinksi (1909, 1912) to describe the climatic and geomorphic conditions of areas peripheral to the Pleistocene ice sheets. It is now used in a more general sense to describe processes and environments in which frost action is dominant. The term cryonival may give a better description of the processes and environments but has not ousted periglacial from general usage. According to French (1976) periglacial environments may be defined as those in which frost-action processes dominate, and the periglacial domain refers to the global extent of these climatic conditions.

Permafrost is a major characteristic of periglacial environments but not exclusively so, although Pewe (1969a) has argued that it is the common denominator of the periglacial environment and is practically ubiquitous in the active periglacial zone. There are many areas which experience frequent freeze-thaw cycles and possess major periglacial landform assemblages but lack permafrost. Taking the broader definition of periglacial, noted above, approximately 25% of the earth's land surface experiences such an environment at the present time. At the height of the Pleistocene glacial periods former periglacial processes had a considerable impact on soils and landforms over a much wider area. Such effects can still be recognized in the soils and continue to effect moisture status and pedological processes. An excellent review of the effect of former periglacial processes and landforms in Britain and Ireland has been provided by Boardman (1987). Because of their importance, periglacial processes have been studied intensively and there is an extensive literature. Good summaries are provided by Pewe (1969a, b), Washburn (1973, 1979), French (1976) and Clark (1988). All that is intended here is to suggest some of the ways in which soils and landforms interact in a periglacial environment.

The soil geomorphology of periglacial environments can be examined in several ways. First, there is the relationship between the nature of soil and superficial materials and their susceptibility to frost action. Fine-grained soils are more susceptible because high-pore pressures can be maintained more easily. Secondly, the availability of water and hydraulic gradients in soils will determine the nature and intensity of frost action. Thirdly, the patterning of soil types in periglacial areas reflects the landforms created.

THE PERIGLACIAL ENVIRONMENT

Low temperatures and the presence of an impermeable, often perennially frozen, substratum are the dominant factors affecting the development of soils in regions where periglacial processes operate. Climatic characteristics vary considerably between continental and maritime locations and between lowlands and highlands. But the dominant characteristics are low temperatures and a varying period of thaw. To give a flavour of climatic characteristics a few areas dominated by periglacial processes are considered. In Svalbard, air temperatures are characterized by relatively high mean values and great temperature fluctuations during winter (Åkerman, 1987; Spinnanger, 1968; Steffensen, 1969). The difference between the highest and lowest monthly means in January and February is about 22°C at Isfjord Radio and Longyearbyen. Isfjord Radio is directly on the west coast of Svalbard at the entrance to Isfjorden, whereas Longyearbyen is further inland along the same fjord and is more continental in its climatic characteristics. Winter temperatures are 2°C lower and summer temperatures about 2°C higher at Longyearbyen than Isfjord Radio. However, all stations possess mean annual air temperatures well below freezing and only in the months of June, July, August and September are monthly mean temperatures above freezing. Svalbard lies well within the zone of continuous permafrost (Liestø, 1977) although it has a milder climate than that generally expected for permafrost regions. Perhaps the best indicator of the intensity and duration of cold temperatures is the freezing degree index (c.f Harris, 1980, 1982). Longyearbyen has a mean annual freezing degree day index of 2467°C days and Isfjord Radio has 1944°C days.

Periglacial environments in the Western Canadian Arctic range from boreal forest and alpine shrub tundra in the south, to tundra and polar desert north of the treeline (French, 1987). Climates range from subarctic continental to high arctic in nature, and summary climatic statistics are shown in Table 11.1. In comparison with Svalbard mean annual temperatures are considerably lower and the winters are more severe; however, summer temperatures are much higher. Two other significant characteristics are highlighted in Table 11.1. These are the small number

Soils and periglacial landforms

of frost-free days and the generally low precipitation amounts, both of rain and snow. In Finnish Lapland, north of the Arctic Circle, the mean date of the first ground frost is 15 October and it thaws between 10 and 20 June (Seppala, 1987). The frost season depends not only on the air temperature but also on the thickness of snow cover and depth of soil. Where snow cover is thin, frost is able to penetrate to a depth of over 3 m in mineral soil while in mires and marshes the average depth of frost penetration is 40–50 cm (Seppala, 1983). Thus climatic characteristics vary, but the dominance of frost action and subsequent thaw produces quite similar responses in the landforms and soils.

Table 11.1 Selected climatic data, Western Canadian Arctic (from French, 1987)

Physiographic Region	Northern Interior Yukon			Mackenzie Delta	High Arctic
Location	Dawson City	Old Crow	Inuvik	Tuktoyaktuk	Resolute Bay
Latitude	64°N	68°N	69°N	69°30′N	74°N
Ecological Region	Northern Boreal	Northern Subarctic	Northern Subarctic	Low Arctic	High Arctic
Elevation (m)	325	245	60	18	64
Mean temperature (°C)					
Annual	– 4.7	– 10.0	– 9.6	– 10.7	– 16.4
January	– 28.6	– 31.7	– 29.0	– 27.2	– 32.6
July	+ 15.5	+ 14.4	+ 13.2	+ 10.3	+ 4.3
Average frost-free period	92	30	45	55	9
Average precipitation					
Annual (mm)	328	203	260	130	136
June–August (mm)	141	96	93	64	69
Snowfall (cm)	132	81	174	56	78

There have been a number of attempts to define and characterize the periglacial environment in climatic terms (e.g. Troll, 1944; Peltier, 1950). Tricart (1970) and Tricart and Cailleux (1967) distinguished three types of climatic periglacial environments. These were:

1. dry climates with severe winters and seasonal and deep freezing;
2. humid climates with severe winters;
3. climates with small annual temperature range with shallow and predominantly diurnal freezing.

Permafrost is characteristic of the first type, is irregular in the second type and absent in the third type. However, there are inconsistencies in the classification. Type 1 includes several different climatic environments and would include the Canadian High Arctic as well as subarctic continental Siberia. A more comprehensive classification is that of French (1976). This is

1. High Arctic climates with a strong seasonal and very weak diurnal pattern producing large annual and small daily temperature ranges. Examples would be Svalbard and the Canadian Arctic.
2. Continental climates in subarctic latitudes with strong seasonal and weak diurnal patterns. There is usually a very large annual temperature range. Examples would include the interior of Alaska and Yukon and Central Siberia.
3. Alpine climates in the mountains of middle latitudes with well-developed seasonal and diurnal patterns. Examples would include the European Alps, the Caucasus, the highest peaks of New Zealand and the mid-latitude areas of the Rocky Mountains.
4. Climates of low annual temperature range in azonal locations and would include islands in subarctic climates such as Jan Mayen, and mountain climates in low latitudes such as the Andean summits and some of the high peaks of East Africa. Permafrost has been noted at 4140 m on Mauna Kea, Hawaii (Woodcock, 1974), in the Andes (Catalano, 1972; Corte, 1978) and on the summit of Citlatepetl in Mexico (Lorenzo, 1969).

Frost action in soils is governed not only by the intensity of the frost but by the number of freeze–thaw cycles. Hewitt (1968) has attempted to differentiate the important phases of the freeze–thaw cycle. Consideration needs to be given to:

1. the absolute or mean number of frost shifts in a given period;
2. the frequency of cycles which also includes the wavelength (e.g. daily, annual) and the recurrence intervals;
3. the intensity of freeze–thaw which depends on the amplitude of temperature change and the slope of the temperature curve;
4. the scale relations of freezing and thawing phases;
5. the problem of what constitutes the 'effective' temperature shift for a given process and the best measure of it.

PERMAFROST

The term permafrost was apparently first used by Muller (1945) to describe the condition of earth materials when the temperature remains below 0°C continuously for some time. The thickness of the permafrost layer is governed by the balance between the internal heat gain with depth and the heat loss from the surface. Above the permafrost is the active layer which thaws during the summer months. The thickness of the active layer is controlled by seasonal temperature fluctuations and many other factors such as vegetation, snow cover, thermal conductivity, aspect and albedo. This is where the nature of the soil and other superficial materials exert an influence. The thermal conductivity of silt is only one-half that of coarse-grained sediments and several times less than rock. The albedo of different soil materials also varies (Brown, 1973), and vegetation is an important factor. Vegetation shields the underlying soil from solar heat during the summer months. French (1976) has argued that the insulating property of vegetation is probably the single most important factor in determining the thickness of the active layer. The active layer is thinnest beneath poorly-drained and vegetated areas and thickest beneath well-drained bare soil and rock. Peat, especially, has a major influence in protecting permafrost from atmospheric heat because of its distinctive thermal conductivity properties (Brown and Williams, 1972; Brown and Pewe, 1973). Dry sphagnum has thermal conductivity values approximately an order of magnitude less than the lowest value for mineral soil.

Landscape topography and aspect also influence the amount of solar radiation received and therefore the freezing and thawing process. The role of aspect in influencing the nature of soil properties has already been discussed. In northern British Columbia and the Yukon, permafrost sometimes occurs on north-facing valley slopes and not on adjacent south-facing slopes (Brown, 1969). Also, the active layer is usually thinner on north-facing slopes but local weather conditions may alter this generalization. Thus, French (1970) has noted that the active layer is thinnest on southwest-facing slopes in the Beaufort Plain of N.W. Banks Island because of the influence of dominant south-west winds. The permafrost zone may contain areas of unfrozen soils called talik. Permafrost is also subdivided into continuous, discontinuous and sporadic permafrost; a zonation which usually reflects distance from the north pole in the northern hemisphere.

SOIL TYPES

The dominant factors affecting the development of soils in periglacial areas are low temperatures and the presence of an impermeable frozen substratum Rieger (1974) even suggests that the cold environment modifies soil-forming processes to such an extent that it overshadows, and may obliterate, the effects of relief and time on soil characteristics. This may be so but it does not mean that within periglacial areas topography or landforms have no influence on the spatial variability of soils. As will be shown later, topography has as great an influence in determining spatial variability of soils in periglacial regions as in any other climatic environment.

A number of workers have argued that soils in periglacial or Arctic areas can be divided into:

1. Poorly-drained soils that are usually, or always, saturated with thin or moderately-thick active layers over solid, ice-rich permafrost.

2. Well-drained soils with dry permafrost that usually have moisture contents below field capacity.

Poorly-drained soils cover most areas of low relief and remain saturated or nearly saturated because permafrost prevents downward water percolation. Lateral movement is also hindered, especially on flat, low-lying areas. Lateral movement can only take place in the thin, upper horizons of soils (Douglas and Tedrow, 1960). Poorly-drained soils consist of a surface organic mat, followed by a grey, greyish-brown or bluish mineral layer which is usually mottled. Streaks of organic matter may occur. The base is usually a grey, frozen layer which may possess patches of organic matter in its upper part, but this sequence may be confused by churning in the soil layers. Warping and heaving producing involutions and pedoturbation is a characteristic of such soils and is a normal factor in their development (Douglas and Tedrow, 1960; Sokolov and Sokolova, 1962). In soils where regular patterns, such as circles, are produced by frost action, buried peat and lenses of organic matter may be forced deep into the mineral material (Ignatenko, 1967; James, 1970). This is portrayed in Figure 11.1. Such processes decline in intensity in regions where thawing is deeper and winter snow cover greater (Ignatenko, 1963).

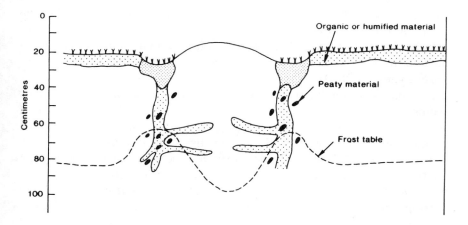

Figure 11.1 Frost circle with lenses of organic material dragged down into lower soil layers (after Ignatenko, 1967).

Such heaving and churning processes and the great variability in thaw and water availability, as well as variable microrelief, produce a soil pattern which can be very complex. The thickness of the organic mat is especially variable. In waterlogged, low-lying areas thick layers of peat may develop and the entire active layer could be formed in organic matter. Much will depend on the nature of the soil materials and the depth of thaw. The texture of soils in periglacial regions is also highly variable reflecting the action of frost-weathering on bedrock and the complexity of superficial materials. However, soils with silty or loamy texture are common. Soil texture as well as climate will affect the depth of thawing. It has been noted that in the eastern European tundra, the permafrost table in poorly-drained soils with loamy textures is often deeper than one metre (Kreida, 1958), but in similar textured soils in northern Siberia it is usually less than 0.5 m deep (Karavayeva and Targul'yan, 1960). Thawing is also influenced by microrelief and topographic position. This is essentially a function of drainage in that sites with good surface drainage thaw to greater depths than those with impeded drainage such as in troughs between polygons (McMillan, 1960; Brown, 1967). The lower portions of solifluction slopes also fall into this category (Smith, 1956).

The nature and distribution of organic matter is also distinctive in Arctic soils. Organic matter tends to accumulate on the surface and is not easily transferred into the soil profile because of an absence of deep root systems. The almost complete absence of vascular plants, low temperatures and precipitation levels also inhibits movement of organic matter into the soil. Organic matter in Arctic soils may not be as obvious as in other soils. Thus Tedrow (1966) has described an upland soil in the polar desert with 6% organic matter in the top 12 cm probably contributed by diatoms and algae but with little apparent darkening. Colourless humus has been noted in Siberian soils (Karavayeva and Targul'yan, 1960). Another characteristic is a zone of organic matter accumulation immediately above and below the permafrost level. This seems more characteristic of shallow soils in areas where the permafrost is close to the surface. It has been suggested that this has been caused by the downward flow of colloids and dissolved humus (Dimo, 1965). This is, in effect, a humus illuvial horizon which may be unique to high Arctic soils. Another possibility is that the layer is fossil and relates to formation during a warmer period (Douglas and Tedrow, 1960). Frost stirring and solifluction may also be responsible for the layer (Mackay *et al.*, 1961; Brown, 1966).

Well-drained soils occur on coarse-textured soils on slopes, narrow ridges, dunes and beach ridges. Under such conditions, water drainage is easier and oxidation rather than reduction can occur. Free water movement also allows the leaching of soluble constituents. Rieger (1966, 1974) has described such a well-drained soil from southwestern Alaska. This soil possessed a dark upper horizon and a yellower subsoil horizon grading into shaly parent material. The soil was acid and low in exchangeable bases. The A_{12} horizon was higher in free iron than the A_{11}

horizon above and the B horizon below. This has also been noted by other workers (e.g. Hill and Tedrow, 1961; Brown and Tedrow, 1964).

Well-drained soils are more stable than poorly-drained soils with less frost heaving and stirring. This results in soils with distinct horizons. However, patterned ground, such as stripes and hummocks do occur on such slopes, and this will introduce some variability. Frost scars, initiated by sorting action, possess thin soils and are affected by processes such as needle-ice formation. Thus, soils in periglacial regions are distinctive because of a combination of climate and geomorphic processes.

MICROMORPHOLOGY OF SOILS IN PERIGLACIAL ENVIRONMENTS

Analysis of soil micromorphology will provide much detailed information about the processes that have been influential in the development of that soil. With respect to periglacial areas, it has been suggested that 'while much remains to be learned concerning the reorganisation of soil materials by periglacial processes, certain micromorphological features have been recognised as characteristic of soils in periglacial environments' (Harris, 1985, p. 219). Four categories of micromorphological phenomena are characteristic of periglacial soils, namely platy or lenticular structure, grain coatings, reoriented skeletal grains and vesicular voids. Many of these can be observed as fossil features in present-day soils (Catt, 1987).

Platy or lenticular structure

Migration of water during soil freezing may cause dehydration of the soil immediately below the freezing front. Frost-heave pressures may also lead to compaction of unfrozen soil. A number of workers have described dense platy peds separated by large planar voids and attributed them to compaction during soil freezing (e.g. O'Brien *et al.*, 1979; Van Vliet-Lanoe and Langohr, 1981). Laterally extensive voids have been described in soils from Arctic Canada by Bunting (1983). These seem to represent the sites of former ice lenses and, because of the pressures created during ice lens formation, often possess thin stress coatings of fine materials. Platy or laminar structures are not necessarily restricted to either permafrost or non-permafrost zones. Pawluk and Brewer (1975) have reported such structures from seasonally frozen soils above permafrost in northern Canada whilst similar structures have been recorded in seasonally frozen soils where permafrost is absent (Federova and Yarilova, 1972). Some workers have suggested that platy structures gives way to a more blocky structure as depth increased or as the permafrost was reached (Van Vliet-Lanoe, 1976; O'Brien *et al.*,

1979). Banded, or plectic, fabrics (Brewer and Pawluk, 1975), are common in the active layers above permafrost. Morozova (1965) has noted them in soils of Yakutia, and Hughes *et al.* (1983) have described them in the Yukon. Such structures also appear common in boreal podzolic soils (Dumanski and St. Arnaud, 1966; Bjorkhem and Jongerius, 1974; McKeague *et al.*, 1974) and even in some chernozem soils where freezing occurs (Coen *et al.*, 1966). The mechanisms involved in the formation of such structures are unclear but probably include compaction and/or dehydration during ice segregation, and possibly an increasing electrical charge associated with flocculating ions in the soil solution during freezing.

Skeletal grain coatings

Plasma coatings on the surfaces of skeletal grains have been known for a long time in soils from Arctic regions. Interestingly they have been described in soils in northern Alaska over permafrost but appear to be rare in soils over permafrost in the Canadian Arctic and Subarctic. They have been ascribed to thixotropic behaviour of the soil mass during thaw and are often associated with patterned ground features (Bunting and Federoff, 1974). Wetting and drying cycles associated with freezing and thawing may also cause skeletal grain coatings. Such coatings appear more common in seasonally frozen soils where permafrost is absent (Bjorkhem and Jongerius, 1974; Federova and Yarilova, 1972; Ellis, 1980a and b, 1983; Harris and Ellis, 1980). It must also be stressed that plasmic coatings on skeletal grains occur in non-periglacial soils, usually due to wetting and drying. But their widespread occurrence in periglacial environments suggests at least a partial contributory cause. The relative abundance in non-permafrost soils compared to permafrost soils appears to be related to soil drainage characteristics. If illuviation of fines is responsible for accumulations on the upper surfaces on coarse grains then rapid vertical drainage is required. In permafrost soils, especially where the active layer is thin and hydraulic gradients are low, free vertical drainage is prevented by the permafrost table. Thus it is difficult for illuviation of fines to occur and for grain coatings to be formed.

Grain coatings may also be caused by gelifluction. Harris (1981a and b) has described smooth-surfaced and streamlined coatings in soils in Okstindan, northern Norway. Rotation of grains in the gelifluction layer led to the coatings. Also plasma may have been plastered onto the grain surfaces during gelifluction to produce streamlined coatings of oriented silt.

Reorientation of skeletal grains

Skeletal grains may become elongated in periglacial soils through the operation of at least two different processes. It has been recognized for some time that stones become vertically oriented because of differential movement of coarse and fine material. Downward penetration of a freezing front results in coarse material moving upwards to the surface and fines moving down in advance of the freezing front. More recently vertical orientation of elongate sand grains has been reported in soils from the Mackenzie River Valley, Canada (Fox and Protz, 1981; Hughes *et al.*, 1983). Fines often concentrate beneath skeletal grains presumably in voids left by melting ice. Many of these characteristics are found in association with patterned ground, thus, earth hummocks often show strong vertical orientation of elongate sand grains (Benedict, 1969).

Other patterns of skeletal grains have been observed such as circular or elliptical arrangements (Federova and Yarilova, 1972; Koniscev *et al.*, 1973). The term orbiculic has been used to describe such patterns (Fox and Protz, 1981). A rather different process has been described by Bunting (1983); he observed ovoid fabrics in wet soils in Arctic Canada and suggested that fines generated by abrasion and/or compression during solifluction formed bands of fine material. The downslope orientation of stones is a major characteristic of solifluction deposits and is used frequently to differentiate solifluction deposits from other deposits. Detailed micromorphological investigations have now established that finer particles can also become oriented. Although, as Harris (1985) points out, evaluation of skeletal grain orientation is complicated by the influence of grain shape on a two-dimensional view.

Vesicular voids

Many arctic and alpine soils, particularly those associated with patterned ground, possess large bubble-like pores or vesicles (Van Vliet-Lanoe, 1985). Vesicles appear to be characteristic of frost circles (Svatkov, 1958; Ugolini, 1966). These may be caused by needle-ice crystals but other origins are possible. As Harris (1985) reports, vesicles have been described in soils from Arctic Canada (Bunting, 1977), Spitzbergen (FitzPatrick, 1956; Chandler, 1972), the South Shetlands (O'Brien *et al.* 1979), Norway (Harris, 1977) and Iceland (Romans *et al.*, 1980). They can be as large as 2–3 mm in diameter, and seem especially characteristic of wet soils. Several explanations have been suggested such as wetting (Romanov, 1974), drying (Bunting, 1977, 1983) and expulsion of air during freezing of wet soils (FitzPatrick, 1956). Thixotropic flow may also generate air bubbles. In a very convincing series of experiments Harris (1983) was able to show that vesicles in soils from the centre of sorted circles and from soliflucted till could be caused by thixotropic behaviour of soils during thaw consolidation. Vesicles can also be created by puddling a wet soil, thus freezing need not be directly involved. However, liquefaction during rapid thaw consolidation is probably significant.

The presence of vesicles in soils in patterned ground implies that lique-faction is an important process in the creation of such patterns.

LANDFORM–SOIL ASSEMBLAGES

The interaction between pedology and geomorphology is best seen in an analysis of the major landform assemblages found in periglacial environments. Landform assemblages can be subdivided into two main groupings: the slope forms created by the processes of erosion and deposition and the various types of patterned ground. Three important slope forms are conspicuous in periglacial areas, all essentially terrace-like features. These are cryopediments, cryoplanation surfaces and soli-fluction terraces.

Cryopediments

Cryopediments are defined as gently-inclined erosional surfaces de-veloped at the foot of valley sides or marginal slopes of geomorphologi-cal units developed by cryogenic processes in periglacial conditions (Czudek and Demek, 1970). Such features range in size from a few metres wide to the massive Old Crow Pediments, covering approximate-ly 50000 km², in the extreme northern Yukon (French, 1987). When they are reasonably extensive and stable it is possible to find simple soil catenary relationships and age relationships similar to those established for pediments in other environments (Chapter 5). However, the specific nature of the soils will be different.

Cryopediments are slopes of transportation, created at the foot of parallel retreating slopes. Processes active on their surfaces are slope-wash, sheetwash and solifluction. The Old Crow Pediments are mantled with a veneer of colluvium and form gently concave slopes, with angles of between 4–9°. As with pediments in semi-arid regions, they possess shallow, integrated drainage systems. Similar surfaces exist in Siberia (Czudek and Demek, 1973). Their similarity with pediments in other environments has prompted French to speculate in the following terms: 'Their great antiquity and their similarity in form with other palaeosur-faces of the world, suggest that long term cryogenic landscape evolution is essentially no different to that of landscape evolution in other semi-arid regions' (French, 1987, p. 41). Soils on such surfaces will be different but the patterning might be similar.

Cryoplanation terraces

Cryoplanation is the name given to the production of an erosion surface by processes involving freeze–thaw activity and periglacial transport processes. Nivation has often been suggested as the mechanism by which the rock scarp retreats to produce the terrace (Reger and Pewe, 1976). Nivation, as a term, was introduced by Matthes (1900) in his study of the glaciation of the Bighorn Mountains, Wyoming and refers to processes associated with snow patches. Thorn (1988) has argued that since then the term has been used indiscriminately. It is generally used to identify the assemblage of weathering and transport processes that are intensified by late-lying snow-patches. Gelifluction and overland flow are also important processes associated with snow-patches and must be considered part of nivation. Thorn (1976, 1979) has shown that transport rates by overland flow are between 20 to 30 times greater within a colluvium-mantled snow-patch site than on a nearby snow-free surface.

Pedologic processes will be related to the activity of the surface and relative position. Soils forming at the foot of an actively developing scarp will be extremely rudimentary and will rely on the production of material by freeze–thaw processes. They will be in a weathering limited situation. An age gradient of soils will exist across the terrace away from the scarp and soils will become better developed and more stable as distance from the scarp increases, depending on the effect of slope-wash processes. Such soils will be developing in transport limited situations. If a suite of terraces exists, soils will be rejuvenated as the lower, retreating scarp is approached. Soils across such surfaces will therefore be diachronous. Also some soils will be undergoing progressive and other soils regressive pedogenesis (Chapter 1).

Solifluction/gelifluction terraces

Solifluction, as a term, is still used to describe the downslope movement, in periglacial areas, of soil saturated with water, Gelifluction, solifluction associated with frozen ground, is probably a better term. Gelifluction is influenced by soil moisture, slope gradient, soil grain-size distribution and vegetation cover. Suitable conditions for gelifluction occur where the downward percolation of water through the soil is limited and where the melting of segregated ice lenses provides excess water which reduces internal friction and cohesion in the soil. Gelifluction can produce a number of small-scale slope features such as lobes and terraces. Such lobes and terraces occur in most periglacial areas but are best developed in areas possessing significant local relative relief. Terraces and lobes vary in size and relief characteristics. Tread angles vary from 3–6° to 20–25° and risers can be greater than 4 m in height. Terraces develop where downslope movement is relatively uniform. The central parts of terraces tend to be wetter than the peripheral areas. This not only affects the nature of downslope movement but also the type of soils. A basic distinction was made earlier between wet and dry Arctic soils. This distinction is often observed on large gelifluction terraces. There may also be a gradation of soils up and down terraces in catena-like fashion. Akin to gelifluction terraces is the gelifluction sheet which can produce extensive low angle (1–3°) landforms. The downslope edge of such sheets is characterized by numerous small lobate forms with Dryas-banked risers, often only a few centimetres high.

Sections through terraces and lobes have revealed buried organic layers which have been overridden by the downslope movement. Such buried layers can be examined to assess rates of soil movement and nature of the previous landsurface (Chapter 12). It may be possible to date this material and to build up a chronology of movement. Soils associated with gelifluction terraces exhibit all the essentials of soil geomorphology. The nature of the soil in the first place will influence the rate and location of movement and the features so produced feed back into the soil-forming processes. The buried soils also allow soil stratigraphic procedures to be used to reconstruct landscapes and assess rates of movement.

PATTERNED GROUND

Probably the most conspicuous features of periglacial environments are the various patterns created by ice and frost action in soils. Such action causes a sorting and redistribution of soil materials, which is conspicuous not only on the ground but persists so that fossil-patterned ground is visible on aerial photographs. This visibility is a function of the soil pattern and the differing moisture characteristics of the sorted materials. The different forms provide information on the lithological, hydrological and climatological situation when they formed and are of value for the reconstruction of past environmental conditions (Akerman, 1987). A variety of patterned ground phenomena exist and only the more conspicuous landforms are discussed here.

Thermokarst

Thermokarst is not usually regarded as a form of patterned ground. Nevertheless it does produce patterns of landforms and associated soils. The term was first used to describe the irregular, hummocky terrain due to the melting of ground ice. However, it is now applied to the process of melting ground ice, including detached masses of glacial ice, irrespective of origin and the landforms resulting from such melting. Thus the ice disintegration landscapes and associated soils described in Chapter 9, would be considered under the general term of thermokarst.

Here, however, the discussion is concentrated on the processes and landforms associated with the melting of ground ice and not stranded glacial ice.

Thermokarst may reflect a long-term regional climatic amelioration or local, non-climatic factors. In a stable climate, thermokarst can develop in response to a variety of geomorphic and/or vegetation conditions. A major natural cause of thermokarst phenomena is the presence of polygonal ice-wedge systems. Water accumulates at the junction of ice wedges in summer and favours more intense thawing. This concentration of water increases as the depression grows larger, promoting what has been called self-developing thermokarst (Aleshinskaya *et al.*, 1972). Thermokarst processes are especially common along the margins of the permafrost zone.

The type of landforms produced and soil materials involved in such processes can be seen by examining a specific example. The type area for thermokarst phenomena is upon the terraces of the Lena and Aldan rivers in central Yakutia. The terraces are underlain by thick sequences of silty loams with abundant segregated ice lenses. Thick ice wedges occur and may underlie over 50% of the terraces. This provides a good basis for thermokarst development, which has been termed alas thermokarst relief and explored by Soloviev (1973).

In stage I, the polygonal system of ice wedges starts to thaw and high-centred polygons begin to develop separated by depressions along the ice wedges (Figure 11.2). Eventually the polygon centres become small conical silty or peaty mounds called baydjarakhs. They are usually 3–4 m in height, 2–15 m wide and up to 30 m long. These mounds then gradually collapse and decay as a depression develops in the centre of the baydjarakh field. Continuous depressions known as dujodas then develop in stage II. By stage III a conspicuous depression, or alas, has formed. An alas is a depression with steep sides and a flat floor with no trees but a thermokarst (thaw) lake. Such a lake deepens and subsurface thawing is enhanced. In later stages (IV) the lake disappears by infilling or drainage. Infilling causes a redistribution and burial of soils as outlined for ice disintegration features in Chapter 9. Accretion gley deposits may form. Because of the drainage of the lake, permafrost may return and new frost heaving and ice wedge formation occur. The end-product is the development of a flat depression with gentle slopes. At each stage of relief inversion material will move off the side slopes and accumulate in the depressions. It may only be the sequence of deposits and buried soils that indicates the evolution of the landscape.

Such a fossil landscape has been described in eastern England (Burton, 1976), where during the Late Devensian, the Fenland was a complex thermokarst terrain. According to Burton (1987) common features of depressions within the Fenland are:

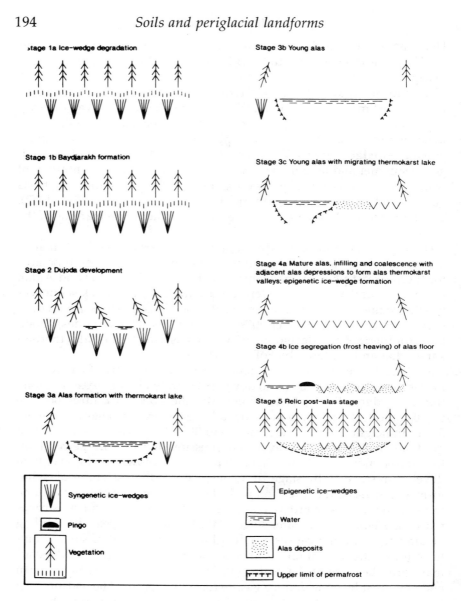

Figure 11.2 Developmental sequence of alas thermokarst relief in central Yakutia (after Soloviev, 1973).

1. the large size, often larger than 1000 m in diameter;
2. enclosed by high ground except for a narrow outlet through which the whole depression drains;
3. bordered by Devensian 1st and 2nd terrace surfaces showing evidence of a polygonal network of ice wedges;

4. excavated through terrace deposits into the underlying Jurassic clays forming flat or gently concave floors;
5. slopes mantled by drift sequences of coarser deposits overlain by finer solifluction material with a loessic component;
6. linked by channels to form alas valley systems.

Pingos

A pingo is an ice-cored hill which has been domed up from beneath by the intrusion of water which then freezes, or by the growth of segregated ice lenses. Pingos are of a larger magnitude than most periglacial features being often 60 m high and 300 m in diameter. Although they represent a classic periglacial phenomenom, 'they should not be regarded as typical features of all periglacial regions. In fact, their existence is usually the result of a number of distinctive and limiting geomorphic and hydrologic conditions' (French, 1976, pp. 93–4). Two types occur; open and closed system types. Open system types occur where permafrost is thin and discontinuous and where water can penetrate the sediments and circulate in unfrozen sediments. When this water rises towards the surface, it often freezes forming localized lenses of ice, forcing the sediments up. They are especially common in the unglaciated areas of Alaska and the Yukon where over 700 have been identified (Holmes *et al.*, 1968; Hughes, 1969).

Closed system pingos occur in areas of permafrost, usually on alluvial plains. They are created by the extension of the permafrost into previously unfrozen areas. The greatest concentration occur in the coastal plain of the Mackenzie Delta, Canada (Mackay, 1962). Mackay (1973) has argued that closed system pingos develop when permafrost extends into a shallow lake basin which is in the process of being infilled or drained. If it is being infilled accretion deposits are likely to result.

Characteristic soil patterns are created on thawing when inversion of relief occurs and raised rims are created. The rims will be quite well-drained whereas the depressions will be waterlogged and gleyed soils will result. Such patterns will remain long after the pingos and permafrost have disappeared. The nature and distribution of ground ice depressions in Britain and Ireland have been reviewed by Bryant and Carpenter (1987) and Coxon and O'Callaghan (1987). The rims vary considerably in height. In Wales, Watson (1971) records a rampart 7 m in height; however, many depressions and rims on the Drenthe loessic plateau of the Netherlands are difficult to detect (De Gans, 1981). Rim or rampart morphology will be inherited from the morphology of the original ice lens. In some cases all the rim material appears to have collapsed back into the basin – such depressions have been observed in East Anglia, England (Sparks *et al.*, 1972). Deformation structures are often found in the sediments of the rampart wall and given the processes involved in the decay stages of pingos, deformation features ought to be present. But, despite extensive excavation, such structures are absent from fossil features in the British Isles (Bryant *et al.*, 1985). In Ireland many pingo remnants are associated with spring lines.

Palsa or palsa-like mounds

A palsa is an ice-cored peat hummock rising from a mire (Seppala, 1972). Palsas range in diameter from a few metres to several tens of metres and in height from less than 1 m to 7 m. They are usually dome-shaped but can form elongated ridges. Palsa complexes, sometimes called peat plateaus, may have diameters of hundreds of metres and flat surfaces. Palsas have characteristic soil structures; the core is mostly composed of segregated ice and frozen peat. There may be a mineral core below the peat cover, with the boundary between peat and minerogenic material being very sharp. Clayey silt with segregated ice is the typical material. Ahman (1976) has noted that the ice content in the upper part of the core is often nearly 100%.

It appears that palsas form best where wind thins the snow cover on a mire allowing frost to penetrate deeply into the peat and causing initial upheaval of the surface. As with all hummock-like features, once the initial microrelief has formed, positive feedback mechanisms operate to enlarge the feature (see earth hummocks). As palsas become larger, they tend to become snow-free and the extent of the frozen layer increases. When the ice lenses melt, inversion of relief occurs producing a low wet centre with raised peat rims. Patterning of soil types will occur accordingly, and several other palsa-like forms occur. Akerman (1987) has described frost blisters, active layer mounds and vegetation peat mounds in Svalbard as well as more typical palsa features. Frost blisters often possess the same size, form and appearance as palsas but generally have steeper sides and cracks occur more frequently on the surface. They appear in slightly different landscape positions being more frequent in marginal parts of mires, along streams, lake and pond shores or near springs. This suggests an association with flowing water and that a groundwater gradient may be essential to their formation. The updoming is the result of a large core of injection ice. The surface peat core is thin, usually less than 30 cm, which accounts for the numerous surface cracks. In many respects they look like mini-pingos.

Active layer mounds are associated with large grass and moss-covered valley bottoms with an ice wedge polygonal pattern. They are smaller than most palsas and their internal structure is different. They are found exclusively in the wet central parts of low-centre ice-wedge polygons and often lack a distinct peat layer. Their internal structure is dominated by a large core of segregation ice situated between a thin vegetation mat and the underlying mineral soil.

Vegetation peat mounds are completely different as frost and ground ice are not responsible for their creation. In size, form and general appearance they look like palsas but they occur in different topographic positions. The main difference is that they occur on dry sites, such as raised beach ridges, and appear to have formed around 'foreign' material such as a bone, a piece of driftwood etc. Akerman (1987) suggests that salts and minerals released have accelerated the growth of mosses and grasses. They also make excellent bird perches and droppings may enhance their formation. When they become a certain size with a sufficient peat thickness, the insulating effect may be sufficient to allow segregated ice to form thus enhancing the mound size.

Earth hummocks

Hummocks, hemispherical or domed non-sorted circles (Washburn, 1956, 1979), are widely distributed in subpolar and alpine environments. They have been described in the Yukon (Sharp, 1942), Greenland (Raup, 1965), Sweden (Lundqvist, 1962) and the USSR (Kachurin, 1959), as well as other areas. They are probably most conspicuous in Iceland where they are called thufa. Such hummocks occur in both permafrost and non-permafrost areas and under forest as well as on open ground.

Earth hummocks occur within quite narrow size limits; widths vary from about 0.4 to 1 m, height 0.2 to 0.8 m and length from 0.5 to 1.7 m. Characteristic dimensions for a sample of Icelandic thufur are shown in Table 11.2. Hummocks seem to develop best on flat to gently sloping surfaces but can occur on quite steep slopes. They also form best on stone-free, fine-grained soils such as aeolian, volcanic, glacial or alluvial deposits. In Iceland, hummock formation appears to have been encouraged by the increase in quantity of windblown deposits as a result of soil erosion (Chapter 12).

Table 11.2 Mean values (metres) of thufur dimensions from a variety of sites in Iceland (from Gerrard, 1992)

	Holokot	Kolgrimstadir	Oxl A	Oxl B	Laufatungur
Width	0.72	0.39	0.92	0.68	0.40
Height	0.22	0.20	0.56	0.29	0.18
Length	0.98	0.53	1.43	1.10	0.53
W/L	0.76	0.77	0.66	0.63	0.76

It is generally agreed that hummocks result from the downward movement of soil in depressions and an upward displacement of mineral soil in the centres. Analysis of the interior of hummocks leaves no doubt that this occurs (Figure 11.3). But how this is achieved is far from certain. In permafrost areas Mackay (1979, 1980) has argued that hummocks develop as a result of alternate contraction and dilation. Freeze–thaw of ice lenses at the top and bottom of the active layer is thought to produce gravity-induced cell-like movement because the top and bottom freeze–thaw zones have opposite curvatures (Figure 11.4). This will not apply to non-permafrost areas but compacted horizons and even the water-table might inhibit downward movement and provide the pressures needed for formation. Once they have formed the processes appear to

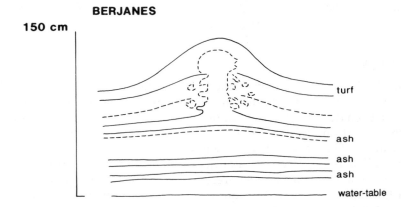

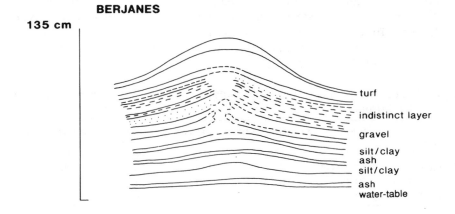

Figure 11.3 Sections through two Icelandic thufa showing disruption caused by differential pressure (from Gerrard, 1992).

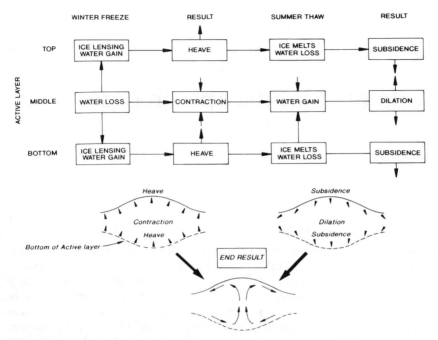

Figure 11.4 Development of earth hummocks by induced cell-like movements (after Mackay, 1980).

be reinforced. Vegetation will develop differently on the mound, possibly insulating the soil underneath. Mounds also are better drained resulting in differential thermal regimes, which also encourages their continued development. Their specific relationships with climatic parameters is uncertain (Gerrard, in prep). Irrespective of their mode of formation they result in microscale differences in soil and vegetation. Soils in hummocks are more porous and better drained; soils in the surrounding depressions are more compact, thinner and occasionally become waterlogged. Microrelief patterns are thus also soil patterns.

CONCLUSIONS

In many respects periglacial regions exhibit landforms on a micro-scale. Nevertheless a close relationship between soils and landforms still exists. It can be argued that most of the principles of soil geomorphology are well-displayed in periglacial areas.

12

Soils and landscape change

GENERAL PRINCIPLES

Soils can provide much information concerning the evolution of landscapes. Soil development takes time, therefore a soil represents a standstill in geomorphological history. The presence of soils may be the only indication that there have been periods of stability within cycles of erosion and deposition. The nature of the soil may also give an indication of environmental conditions during the period of soil formation. Calcium carbonate concretions, iron pans, clays such as kaolinite and various exchangeable cations can provide valuable insights into specific environmental conditions. However, correlations between soil properties and environments are not sufficiently established to permit unequivocal conclusions to be drawn, but the

> history of depositional and erosional events and of soil-forming processes in various environments that together produced the soil mantle of an area can be reconstructed by detailed field studies with supporting petrographic, chemical and other analyses of representative profiles and of unaltered rocks or unconsolidated sediments thought to resemble the parent materials of important soil horizons (Catt, 1986, p. 167–8).

Soils of greatest use in this respect are those which have been buried under later deposits and have had their characteristics essentially fossilized. It is for this reason that buried soils have a vital role to play in Quaternary sciences, since they frequently provide the only record of time breaks within the stratigraphic record. It is necessary, when using soils as stratigraphic tools, that features of the soil profile that are mainly sedimentological in origin can be distinguished from those whose origin is pedological. This is the same problem as that encountered in an examination of alluvial soils (Chapter 6). This uncertainty means that soils should not be used solely, and to the exclusion, of other criteria. A stratigraphical approach to the study of soils should begin with the materials observable in the field leading to an investigation of processes, and perhaps, concluding with some account of the history of soil development. However, the last stage is not without its problems, as later sections will show. A stratigraphical approach also facilitates the coordination and synthesis of information from related fields of the Earth sciences.

Soils used for correlation purposes are called soil stratigraphic units. The American Stratigraphic Code (American Commission on Stratigraphic Nomenclature, Article 81, 1961), defines a soil stratigraphic unit as a soil with physical features and stratigraphic relations that permit its consistent recognition and mapping as a stratigraphic unit. As stratigraphic relations are the crucial elements, any soil stratigraphic unit may, and probably will, consist of several distinct pedologic units. The soil should be studied as a mantle; a mantle that can be traced over wide areas. The pedological properties may vary but the stratigraphic relationships are constant. To be efficient as a soil stratigraphic unit a soil must have features that are pedologic in origin and that display a consistent relationship to other units in the stratigraphic succession. For this reason it is more realistic to think in terms of type transects rather than type sites. Thus, one of the conclusions of the INQUA Commission of Paleopedology (Yaalon, 1971) was that profiles should be traced laterally to determine their spatial variation. This plea for a type transect has also been made by Follmer (1978) for the Sangamon soil of North America, which is probably the most widely and best studied of all buried soils. This transect would be, in essence, a palaeocatena and should contain the complete range of drainage and topographic conditions common to the soil stratigraphic unit. In the case of the Sangamon soil, this should include a poorly-drained, organic-rich accretion gley profile, a poorly-drained *in situ* profile, an imperfectly-drained profile and a well-drained profile.

Correlations over short distances are essential to a detailed reconstruction of landscape development. This is achieved largely by the analysis and interpretation of buried soils. Morrison (1967) proposed the term 'geosol' for the basic soil stratigraphic unit but it has yet to be universally accepted. The widely-used term palaeosol is the one adopted here. There is a considerable literature on palaeosols and all that is attempted in this chapter is to highlight some of the methodological problems involved in their identification and interpretation. More comprehensive treatments have been provided by Morrison and Wright (1967), Yaalon (1971), Mahaney (1978a), Boardman (1985a), Catt (1986) and Retallack (1990), while Valentine and Dalrymple (1976) have produced an excellent summary of Quaternary buried palaeosols.

Pawluk (1978) has argued very strongly that emphasis in the study of buried soils should be placed on the pedogenic rather than the pedologic profile. The pedogenic profile consists of the interactions within and among the processes contributing to the dynamics of the soil body. Concentration on the pedogenic profile allows palaeoenvironmental conditions to be based on energy relationships rather than matter alterations. Since several morphologically different soils may form within the same pedogenic setting, because of material and other differences, concentration on the pedogenic setting is less prone to error. Evidence for processes rather than effects should be sought.

The great variability in pedological properties makes long-distance correlation difficult and perhaps unrealistic. Changes in energy factors, materials and geomorphological frameworks over even relatively short distances, makes it imperative that as many sections as possible are analyzed. Thus, Richmond (1962) has shown how, in Utah, a buried brown podzolic soil at high levels changes through a brown forest soil to a sierozem at lower levels. Well-developed buried peat beds may grade gradually into organic silts and then to purely mineral deposits over distances as short as a hundred metres. The stratigraphical relations are constant but there has been a gradual facies change and genetically different soils can occur at different levels. Thus, the suggestion by Firman (1968) that soil stratigraphic units may be traced across soil and climatic zonal boundaries, poses numerous problems. Likewise, transcontinental correlations based on soils with the greatest development, such as that by Richmond (1970), between the interglacial deposits of the Rocky Mountains, the Sangamon palaeosol of mid-west North America and similar deposits in Europe are open to doubt. Some of the techniques employed in making these correlations are now examined.

TYPES AND RELATIONSHIPS OF PALAEOSOLS

A palaeosol is a soil which has been formed on a past landscape. This landscape may be buried or exposed. Buried palaeosols occur where the land surface has been covered by younger deposits, whereas relict palaeosols occur on surfaces that have never been buried. Palaeosols formerly buried but now re-exposed by erosion are termed exhumed palaeosols. Buried palaeosols have their characteristics largely fossilized by the surface accumulation of various sediments. Some secondary changes may occur, such as deposition of iron, manganese or calcium carbonate, and organic matter may be oxidized, but the changes are normally minor. Several potential relationships between buried, relict and modern soils are possible. Catt (1986) has questioned the use of the term palaeosol, as defined above. He queries whether 'landscape of the past' should include all buried surfaces, even those buried extremely recently, such as by last year's hillwash, or only surfaces dating from past periods when environmental conditions were sufficiently different to produce detectably different soil features. If the former definition is accepted, the unburied equivalent could be called a relict palaeosol and all soils become palaeosols. If the latter definition is adhered to, the age distinction between palaeosols and younger soils becomes blurred and varies widely. Catt (1979) has also stressed that the majority of soils contain elements which formed as a result of environmental conditions which have now altered. If the detailed stratigraphic relationships of the soil is described Catt (1986) is right when he argues that there is no need to use the term palaeosol. The terms buried soil, relict soil and modern soil are probably sufficient.

A weakly-developed relict soil will have its characteristics altered by a subsequent stronger pedogenesis. Therefore, weakly-developed soils may only be identified when they occur beneath younger deposits. As they reach and merge with the surface their characteristics will be obliterated by a stronger soil (Figure 12.1(a)). Alternatively, if the buried soil is strongly developed, pedogenesis during later soil-forming intervals may be unable to alter its characteristics. The situation displayed in Figure 12.1(b) will then arise. Combinations of these possibilities produce complex relationships (Figure 12.1(c)).

Other possible relationships exist and have been termed composite, compound and subdivided soils (Morrison, 1978). Detailed examination of the soils might provide evidence of more than one episode of pedogenesis within the same profile. Such a soil is composite (Figure 12.2(a)). The sequence of events can be determined but the length of time between pedogenic events is indeterminate, although the stratigraphic relations between the various events may be discovered by tracing the soil to sites where the soil-forming events are clearly separate. In Quaternary stratigraphy it was commonly assumed that one soil was associated with one interglacial, but it is now clear that several phases of soil development can occur in the same interglacial. Relict soils are always polygenetic to some degree but many buried soils are also polygenetic. Good examples are the Paudorf soil of Austria and the Sangamon soil of North America.

Composite soils can be produced by the process known as soil welding (Ruhe and Olson, 1980) which is the formation of the solum of a ground soil through a thin cover sediment and mergence with the solum of a buried soil formed in a substratum material. The sequence begins with a simple A/B soil profile. The soil is then eroded, deposition of new material occurs on this truncated soil and a younger soil forms in the overlying material and develops in the remains of the older soil. The textural difference at the interface of the two soils interferes with vertical water movement, the soil-water chemistry is changed and the precipi-

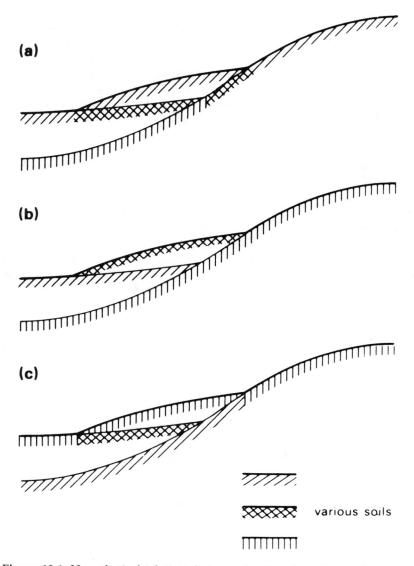

Figure 12.1 Hypothetical relations between fossil and modern soils.

tation of excess ions in solution may occur. Induration then begins. Soil welding complicates recognition of boundaries between materials and stratigraphically aligned soils. The boundaries can usually be identified using physical and mineralogic properties. Chemical properties, in general, are not very useful because the buried soil commonly is secondarily enriched with chemical components brought down from the overlying soil.

A   horizon of both soils

composite B horizon

Cca horizon of younger palaeosol

B   horizon of older palaeosol

Cca horizon of older palaeosol

(a)

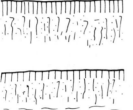

younger palaeosol

older palaeosol

(b)

Figure 12.2 (a) Composite and (b) compound palaeosols (from Morrison, 1978).

Multi-storey soil profiles are those where more than one soil profile occurs in close vertical succession, but are separated by sediment, so that the individual soil profiles do not overlap. These have been called compound soils (Figure 12.2(b)). Soils that usually occur as single composite soils may be traceable to places where they are separated by sediment into two or more soil profiles (Figure 12.3); these are called subdivided soils. It is clear that what one worker calls a subdivided soil may be described by another worker, in a different area, as separate soils. Morrison (1978) has suggested that this has been so with the Paudorf soil.

post-palaeosol deposits

subdivided palaeosol

Figure 12.3 Subdivided occurrences of a single palaeosol (from Morrison, 1978).

This soil was thought to represent a Late-Middle Wurm interstadial but it has now been shown to represent a time span from the start of the last Riss-Würm interglacial to the beginning of the main Würm glacial stage. Subdivided equivalents of the Paudorf soil are common in Czechoslovakia.

The similarity between buried and relict soils presents many problems. Birkeland (1974) suggests two mechanisms whereby this similarity may be achieved. During periods of fluctuating climate, soil formation could occur in a stepwise fashion, where rapid development takes place under optimum climatic conditions. A curve of soil development with time such as that depicted in Figure 12.4 (curve A) would be produced. Climatic change at time I retarded soil development and surface processes buried part of the pre-existing soil. Because little soil development takes place between I and II, the surface soil that formed between 0 and II will be similar to the soil that formed between 0 and I but which has been buried between I and II. Soil development during time interval II–III should produce a more strongly-developed soil profile than in the buried profile. Another possible explanation for the similarity between surface and buried soils is that both have reached a steady state in development so that little further change is possible (Figure 12.4, curve B). In areas where major climatic fluctuations have been frequent, this explanation is less likely because, although some soil properties seem to achieve a steady state relatively quickly, others need longer time periods. The first explanation accords with the view that appreciable soil development only occurs during interglacials and is considerably inhibited in glacial times. But evidence from high Arctic and Alpine areas shows that soil formation is far from inhibited. Therefore, the flattening

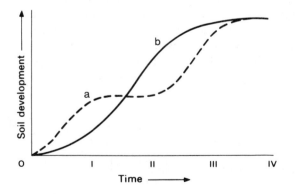

Figure 12.4 Hypothetical curves for soil development under (a) a fluctuating climate and (b) a constant climate. (From *Pedology, weathering and geomorphological research*, by Peter W. Birkeland (1974). Copyright 1974 by Oxford University Press Inc. Reprinted by permission).

off of curve A (Figure 12.4) is too pronounced and a considerable dilemma still exists. Another major problem is that comparison between surface and buried soils should be between soils from similar topographic positions.

The mode of burial is often overlooked when assessing the pedological features of buried soils. In many instances burial is quite sudden, such as by slope or glacial deposits. But the possibility that the soil and vegetation can continue to develop and incorporate the burying medium must not be overlooked. Many buried soils occur in loess deposits and it is conceivable that soil and vegetation would not be inhibited unless the rate of deposition was particularly great. The interpretation of fossil soils is, therefore, still beset with numerous problems.

IDENTIFICATION OF BURIED SOILS

The unequivocal identification of soils in deposits of similar composition is difficult to establish on field evidence alone and one is forced to agree with Ruellan (1971) that the identification of a buried soil in a single section is rarely simple. Valentine and Dalrymple (1976) argue that some of the difficulties stem from the looseness with which terms such as soil, regolith and weathering zones are used. But it is the nature of the soils themselves that produces the greatest problems, namely, the ephemeral nature and susceptibility to change of many soil features and the similarity of many soils to sediments. In many buried soils the A horizon may be absent because of erosion, yet it is largely the properties of the A horizon that are important in soil classification. The problem was encountered by Bruce (1973) in New Zealand loessic deposits. The A and B horizons had been stripped off prior to the deposition of the succeeding loess and the morphological structure of the majority of the buried soils was comparable to that of the subsolum horizons of the present soils. Good summaries of the problems of recognition and interpretation of buried and relict soils have been provided by Fenwick (1980, 1985) and Jenkins (1985).

A great many techniques can be employed in the identification of buried soils as indicated in the report of the Working Group on the Origin and Nature of Palaeosols (1971). A simple division can be made between field techniques utilizing observable characteristics and more detailed laboratory techniques. Accurate assessment requires a combination of both approaches. Colour has been used by many workers, e.g. Butler (1958) in New Zealand and Kobayashi (1965) in Japan. But colour can be deceptive and colour changes can occur with comparative rapidity. Raeside (1964) has used colour but in combination with structural features, root traces and clay skins. Various morphological characteristics have also been used; Simonson (1954) employed a combination of structure, cementation and clay sesquioxide accumulation. Decalcification has been one of the most commonly used field indicators and has been used extensively to differentiate till sheets in North America. Gile and Hawley (1966) relied on the evidence of decalcification in differentiating soils on alluvial fan environments in New Mexico.

Laboratory techniques are being employed increasingly to identify buried soils. Studies of compounds in which phosphorus occurs in soils indicate that the proportions occurring in different modern soils are characteristic of these soils and reflect their history and genesis. The relative proportions of primary, residual and organic phosphorus compounds were employed by Leamy and Burke (1973) to identify buried soils in Central Otago, New Zealand. These authors conclude that the field identification of such soils can be confirmed by identifying and measuring the forms of phosphorus present. Also, the interpretation of the relative proportions of the phosphorus fractions provides a technique for assessing the relative degrees of weathering and possibly the length of time of soil formation. Goh (1972) and Thompson *et al.* (1981) identified buried soils by using the level of amino-acid nitrogen as an indication of biological activity. Limmer and Wilson (1980) found that the number of amino acids in New Zealand buried soils decreased from 14 to 6 with increasing age up to 41 000 years. Childs (1973) measured the vertical variation in total amounts of specific elements to detect the redistribution caused by soil processes, and Birrell and Pullar (1973) determined the amounts of elements such as aluminium and iron present in an extractable form as a measure of soil development. Infrared spectra of humic acids in buried Ah horizons (Dormaar, 1967; Reeves and Dormaar, 1972) have been used to distinguish between forest and grassland soils and between the B horizons of buried chernozems and gley soils (Dormaar, 1973). In addition, various biological indicators such as pollen (Dimbleby, 1952), mollusca (Kerney, 1963), beetles and opal phytoliths (Dormaar and Lutwick, 1969) aid the identification of buried soils.

Many soils show characteristic changes of composition with depth from the surface. Such depth functions, in combination with other features, may be able to identify buried soils or to distinguish soil material *in situ* from transported soil materials. Bateman and Catt (1985) have used the depth at which 90% of the maximum amount of each mineral in the soil profile has been removed (the D value). For similar types of soils the D values for individual minerals should occur in the same order. Depth functions of micromorphological features have been used extensively (e.g. Brewer, 1972; Finkl and Gilkes, 1976; Kwad and Mucher, 1977, 1979; Bullock, 1985; Bullock and Murphy, 1979). Catt (1986) has suggested that the most useful micromorphological features to confirm the identification of a buried soil are depth functions of argillans and other coatings (e.g. ferrans, calcans), pores, papules, ferruginous and manganiferous concentrations, secondary minerals, plasmic fabric types, faecal pellets, components of indigenous biological origin (e.g. phytoliths, diatoms) and trace fossils, earthworms or root channels.

PROBLEMS OF ENVIRONMENTAL RECONSTRUCTION

Features which may be useful in recognizing buried soils may be meaningless for an accurate palaeoenvironmental reconstruction. The identification of buried and relict soils is very important in indicating changes in geomorphological systems, but it is even more important to be able to come to some conclusion concerning the climatic and vegetational environment represented by the soils. But this step is fraught with many problems. It was stressed in Chapter 1 that detailed quantitative relationships between climate and soils have still to be established. It is generally assumed that soils which formed under similar climatic, biotic and drainage conditions during similar time intervals are similar. It is also assumed that the development of soils is a function of time in an absolute sense, such that weakly-developed soils must be young and well-developed soils must be old. But these are assumptions that might not always be true. There is also a great danger in using simple modern analogues to explain features of buried soils. Colour is one of these. There is a temptation to equate soils possessing deep-reddish horizons with modern analogues that are developing in warm, humid climates. Pawluk (1978) has argued that the presence or absence of characteristics which reflect a high degree of weathering, may not result from the kinds of processes but rather from the length of time during which the processes have been active. Greater significance should be attached to those processes in which a threshold level of activation energy must be achieved for a reaction to take place. Identification of these processes in buried soils may reveal much more about the environment of formation. This dilemma can be illustrated in the work of Rutter *et al.* (1978) in the Central Yukon, North America. Evidence for differences in climate between two interglacials was obtained from the type of clay minerals and their depths in two buried soils. Montmorillonite-kaolinite mixed clay minerals were found at depths of 190 cm in one whereas vermiculite-chlorite intergrades were found at depths up to 93 cm in another. Using a formula developed by Birkeland (1974), it is possible to estimate the amount of water needed to reach the two depths mentioned. This indicated that the climate in which the first soil developed was probably considerably more humid than for the second soil. But the differences in depths of water penetration may also be a function of time. The nature of the clay minerals may be more indicative of climate, the inference being that the climate was warm and dry to produce montmorillonite.

The identification of biological components may be more useful in suggesting specific environments. Organic matter type and content and carbon : nitrogen ratios have been used. If the assumption can be made that the initial composition of the plants and organisms making up the bulk of the raw organic matter is relatively similar, then an evaluation of the persistence of organic fractions with different stabilities could provide a useful measure of the environmental conditions under which organic matter decomposition took place (Pawluk, 1978). This procedure is complicated by further degradation after burial. The more readily decomposable fractions, such as loosely-bound proteins and carbohydrates, are removed whereas the more resistant constituents, such as ligno-proteins, may be enriched. The work of Bal (1973) in this respect is extremely exciting. It is suggested that the decomposition and distribution of organic matter are related and can be regarded as an individual entity for which the name 'humon' is suggested. Humon is a fingerprint for the soil body. It is defined as a collection of macroscopically and microscopically observable organic bodies characterized by a specific morphology and spatial arrangement. Original plant components may be recognized by the identification of cellular structures and their decomposition products. The recognition of resistant organic bodies also allows original populations of plants and animals and their associated environments to be discerned.

As mentioned earlier, infrared absorption spectra of humic acids can be used to recognize soils. The method may also allow the type of environment to be specified. Thus, Dormaar and his associates were able to differentiate forested soils from grassland soils by this method (Dormaar and Lutwick, 1969; Reeves and Dormaar, 1972). Pawluk (1978) employed the same technique in a comparison of a buried black chernozemic soil in Edmonton, Canada, with modern soils. The slopes on the infrared spectra for both soils confirmed the presence of grassland vegetation but a more pronounced slope for the humic acid extracted from the buried soil was believed to reflect the removal of low molecular weight constituents through post-burial degradation. A major problem with this type of analysis is that A horizons of buried soils which contain the majority of the organic matter have been removed, nevertheless it has been used successfully on the B horizon of truncated palaeosols (Dormaar, 1973).

Buried soils may also be dated by the thermoluminescence method (Wintle, 1981; Wintle and Catt, 1985; Wintle and Huntley, 1982; Wintle *et al.*, 1984). But the method requires that a soil's thermoluminescence be removed (i.e. bleached) by sunlight prior to burial, thus it is best used on buried A horizons. One major problem is that horizons that have been gleyed or decalcified give thermoluminescence dates younger than those of unaltered parent material. However, the age range over which the method can be used is greater than that for radiocarbon dating and the age obtained from an A horizon provides a date for burial without any residence time error.

Biological indicators such as faecal pellets and pedotubules may allow specific faunal species to be identified. Similarly sclerotia and spores indicate the presence of fungi; phytoliths, pollen grains and plant macrofossils indicate vegetation characteristics; diatoms may suggest wet conditions and the resistant chitinous exoskeletons of beetles allow species identification and a picture of palaeoenvironments to emerge. A variety of techniques is available and should be used in combination, but each possesses particular problems. Some of the more important of these techniques are now discussed.

Many of these indicators require that biological components remain fossilized in palaeosols. Thus bones are not preserved in acid palaeosols or leaves on oxidized profiles. As Retallack (1990) has pointed out, numerous processes act to affect the fossil record of palaeosols. Water-logged soils are most likely to preserve leaves and pollen. Spores and pollen may be preserved in well-drained soils with a pH less than 4 or greater than 9. In the Early Eocene Willwood Formation of Wyoming, fossil snails and mammal bones are found mainly in well-drained red Alfisols formed under stable floodplain forests (Brown and Kraus, 1981). Retallack (1984) has produced a stability model based on pH-Eh relationships and the moisture status of the soil.

WEATHERING INDICES

The intensity of chemical weathering that soils have undergone can be assessed by constructing ratios of chemical compounds (chemical weathering indices) or minerals (mineral weathering indices). Many of the chemical weathering indices employ the amount of silica (SiO) or alumina (Al_2O) or both in soils. Many alkaline earths are removed more readily than alumina, and silica is particularly resistant to chemical decomposition. The more commonly used indices are listed in Table 12.1. The molar ratio $CaO : ZrO$ can also be used because the calcium-bearing mineral hornblende is more susceptible to weathering than the more resistant zircon. As an example, silica : sesquioxide ratios and potassic : silica ratios were used by Rutter *et al.* (1978) to differentiate three palaeosols in the Central Yukon. The potassic : silica ratios are shown in Table 12.2.

The same principles underlie the use of mineral weathering indices. Goldich (1938), after studying the persistence of minerals in soils following weathering, proposed a stability series for the commonest minerals. This information is utilized in constructing mineral weathering indices. One of the most widely used is the ratio of quartz to feldspar; the higher the ratio the greater the amount of chemical weathering. Some indices employ heavy minerals and rely on the stability series developed by many workers (e.g. Pettijohn, 1941; Dryden and Dryden, 1946). Olivine, amphiboles and pyroxenes are least stable; zircon and tourmaline are most stable. Thus, the ratio of zircon and tourmaline to the amphiboles and pyroxenes provides a useful indication of the stage reached by chemical weathering. Ruhe (1956) employed both these indices to good effect in the comparison of soils in Iowa (Chapter 6). This work emphasizes that the state of weathering will vary within soils depending on the horizon sampled and, therefore, great care must be taken when comparing palaeosols where horizons may be difficult to identify. Also, more than one particle-size fraction must be examined because mineral species content varies with particle size. These indices may be reinforced by considering the stability series of weathering of clay and clay-sized minerals. The most susceptible to weathering are gypsum, calcite, olivine and biotite and the least susceptible are anatase, haematite, gibbsite, kaolinite and montmorillonite (Jackson *et al.*, 1948). These techniques rely on accurate information concerning soil formation and the sequence of weathering changes. The efforts by Evans (1978) and McKeague *et al.* (1978) to quantify the changes that take place during pedogenesis are invaluable in this respect.

Table 12.1 The more commonly used chemical weathering indices

$\dfrac{SiO_2}{Al_2O_3}$	silica : alumina ratio
$\dfrac{SiO_2}{Fe_2O_3}$	Silica : ferric oxide ratio
$\dfrac{SiO_2}{Al_2O_3 + Fe_2O_3}$ or $\dfrac{SiO_2}{R_2O_3}$	silica : sequioxide ratio
$\dfrac{K_2O + Na_2O}{Al_2O_3}$	alkali : alumina ratio
$\dfrac{CaO + MgO}{Al_2O_3}$	alkali earth : alumina ratio
$\dfrac{CaO}{MgO}$	calcic : magnesia ratio
$\dfrac{K_2O}{Na_2O}$	potassic : sodic ratio
$\dfrac{K_2O}{SiO_2}$	potassic : silica ratio

Table 12.2 Potassic : silica ratios of Yukon palaeosols (from Rutter *et al.*, 1978)

Pre-Reid (early Pleistocene)		Reid (Illinonian or early Wisconsinian)		McConnell (Wisconsinian)	
Horizon	K_2O/SiO_2 $\times 10^{-2}$	Horizon	K_2O/SiO_2 $\times 10^{-2}$	Horizon	K_2O/SiO_2 $\times 10^{-2}$
II Btl	2.5	II Bml	4.2		
II B2	1.7	II Bm2	4.3	II Bm	4.1
II BC	2.0	II BC	4.3	II C	5.0
II C	2.0	II Ck	5.0	III C	5.1

RADIOCARBON DATING

The radiocarbon method is the major dating method applicable to soil and superficial sediments and has been used extensively (e.g. Scharpenseel, 1971; Scharpenseel and Schiffman, 1977; Matthews, 1980, 1981, 1984, 1985; Matthews and Dresser, 1983; Ellis and Matthews, 1984). But the method can only date material up to about 40 000 years old. Materials suitable for radiocarbon dating include peat, wood, charcoal, organic mud, soil humus and calcium carbonate in molluscs and bones; inorganic carbonates can also be dated. Cosmic rays produce radioactive ^{14}C by bombarding ^{14}N atoms in the atmosphere. The ^{14}C combines with oxygen, producing carbon dioxide, which is taken up by plants in photosynthesis. Animals become radioactive by eating plants. When the organism dies, radioactive decay begins and half the specific radioactivity is lost after a period of time known as the half-life. Libby (1955) calculated this as 5570 ± 30 years. Subsequently the half-life was recalculated as 5730 ± 40 years, but to avoid confusion radiocarbon dates are still published with respect to the originally calculated half-life. Thus, the sample can be dated by determining its specific activity.

The dating of palaeosols can be attempted for both the organic and inorganic carbon in the soils. The dating of soil organic matter is complicated by the fact that the different humus fractions, namely humin, humic acid and fulvic acid, will have different ages. Thus, the true age of the soil is impossible to determine and the date obtained represents the mean residence time of the various organic fractions. For palaeosols this is the time since burial (Campbell *et al.*, 1967). Organic fractions dated for modern soils in Saskatchewan, Canada, showed humic acids as 1308 ± 64 years, fulvic acids as 630 ± 60 years and humin as 240 ± 60 years old (Pawluk, 1978). Minor environmental changes, such as topography and drainage, can produce considerable differences in dates by controlling the addition of small amounts of fresh organic matter. Contamination, producing erroneous dates, is possible by further carbon exchange with the atmosphere or with groundwater and through the mixing of materials of different ages.

Also, more recent carbon may be added to buried soils by the filtration of humic acids and, perhaps, by deep root penetration. As Shotton (1967) and Olsson (1974) show, only a small amount of contamination is necessary to affect considerably the date obtained.

The variability of dates within palaeosols may be large and the question of some contamination must always be considered. Ruhe (1975) has described a palaeosol from Australia where hand-separated carbonized specks were 33 700 ± 2230/1730 years old, a fine-earth fraction containing organic carbon was 19 980 ± 370 years old with its NaOH soluble fraction 24 960 ± 580 years old and its NaOH insoluble fraction 25 360 ± 580 years old.

The dating of inorganic carbonates also poses problems (Bowler and Polach, 1971). In dry environments inorganic carbonate dates are often older than they should be, whereas in moist climates, post-pedogenic exchange of modern ^{14}C gives younger dates. Nevertheless, if the technique is applied carefully realistic dates for palaeosols can be obtained. Other techniques are now being used to supplement and extend the radiocarbon dates. Soviet workers studying loess in Tajikistan have been reasonably successful in extending dates back to 2.5 million years by thermoluminescence and palaeomagnetic methods.

POLLEN ANALYSIS

The use and interpretation of pollen diagrams is standard practice in the environmental sciences and the methodology has become well established (Faegri and Iversen, 1974). Pollen may be preserved in acid soils for long periods and the distribution of the various pollen types will give a general picture of past vegetational changes. But analysis of pollen from soils should not be viewed as pollen analysis of bog or peat sites. Similarly, pollen diagrams from buried peat deposits should be interpreted differently from those of buried soils. Most pollen in peat samples is transported from nearby trees; little comes from vegetation growing on the site and little has been transported in from elsewhere. Pollen analysis of soils provides a site or local picture of the regional pollen rain (Dimbleby, 1961a).

All pollen, not only tree pollen, is counted and represented in soils. Thus, the predominance of non-tree pollen is a sure indication of the absence of trees. Also, the flowering of plants under trees is reduced to an extent that they represent only a small proportion of the total pollen. A mixture of non-tree and tree species suggests a mosaic of vegetation types as the trees must be near enough to contribute to the pollen rain.

Pollen analysis of buried palaeosols permits statements to be made of the vegetation at the time of soil formation (Caseldine, 1983). Analysis of surface, apparently modern, soils has also provided extremely useful information about surface processes. Investigations on many soils by

Dimbleby (1961b) have shown that about one-third of the soils sampled contained old soil levels, more or less disturbed, buried beneath deposits of soil and gravel. Pollen is normally abundant at the soil surface and the amount decreases rapidly with depth. Old surfaces show in pollen analysis as sharp breaks in the curve due to the difference in pollen content. They may also be represented by a change in pollen composition. In many cases the boundary between transported soil and *in situ* material cannot be seen in the soil profile and it is only the pollen that indicates the change. These have been called buried levels by Dimbleby (1961b) rather than buried surfaces or buried soils. Soil creep is the most likely process involved in the movement of the transported material and it clearly leads to unusually thick A2 horizons. This thickness has been created either by a new A2 horizon developing on the transported material or the original A2 horizon extending in depth as new material is added. The rate of transportation is, therefore, of prime importance. This evidence also implies that simple statements about landscape and soil stability should be avoided and that pollen analysis deserves a wider application in both contemporary and fossil-soil investigations.

TEPHROCHRONOLOGY

In many volcanic areas such as New Zealand, Iceland, South and Central America, periodic ash or tephra falls mantle the surface. Tephra has been defined as all the clastic volcanic materials which, during an eruption, are transported from the crater through the air (Thorarinsson, 1954). The colour and petrology of the tephra are often characteristic of a particular volcano and can be dated with reasonable accuracy. Thus, many of the tephra layers mantling the surface of Iceland have been dated (Thorarinsson, 1944; Larsen and Thorarinsson, 1978). Rhyolitic tephra eruptions tend to be paroxysmal with long periods of time separating bursts of volcanic activity. This may allow soils to form between each eruption which will be buried by the next tephra fall. But, many tephra falls are not sufficient to kill surface vegetation and are incorporated in the developing soil. Much will depend on the intensity of the eruption, the distance from the volcano and the wind strength and direction at the time of the eruption.

Typical deposits in these volcanic areas show a succession of ashes, mineral and organic layers. Some of the intervening layers may be palaeosols. The ash stratigraphy is, therefore, a great help in elucidating the sequence of palaeosols. Geomorphologically, the ash layers allow interpretation of the erosion and deposition that occurred between falls. The ash layers represent isochronous surfaces and the relative thickness of material between similar ashes at different localities enables tentative conclusions to be reached concerning landscape history. Also, as the ash layers mantle the pre-existing surface, former erosional features may be preserved.

OPAL PHYTOLITHS AS ENVIRONMENTAL INDICATORS

Some of the silica absorbed by plants is precipitated in plant cells to form opaline substances known as opal phytoliths. These possess characteristic shapes and sizes and, when the plant decays, they remain in the soil. Some soils are as much as 60% phytoliths near the surface (Dixon and Weed, 1977). Grasses are the highest producers and yield phytoliths that can be distinguished from most other plants, although grass phytoliths are very similar to those produced by rushes and sedges. Modern forest and grassland soils differ in their phytolith contents and this difference can be used to infer gross vegetational changes (Jones and Beavers, 1964). In effect, they can be used to detect relict palaeosols.

Phytoliths also persist for long periods of time and, therefore, will be present in buried palaeosols (Wilding, 1967). However, plant opal is more soluble than quartz at a pH of less than 9 and both are dissolved at higher pHs (Leo and Barghoorn, 1976). Phytoliths can provide important information on vegetational changes. In this way, counts of opal phytoliths have helped to explain the origin of palaeosols in several river valleys in southern Alberta (Dormaar and Lutwick, 1969). However, data on the rates of production and destruction of opal are required for really meaningful interpretations, since some silica may be easily available from the weathering of volcanic glass. The influence of climate and soil conditions on the weathering of opal is also largely unknown.

RECONSTRUCTION OF PAST LANDSCAPES

Studies of buried soils provide strong evidence of repeated phases of slope instability in many parts of the world. Some of these phases are related to fluctuations in climate during the Pleistocene but they are also related to local environmental changes of which stripping of the vegetation and overgrazing are possible causes. The combination of the techniques and stratigraphic principles outlined in this chapter provides a powerful research tool with which to elucidate some of these changes. A few examples are now examined to demonstrate how these principles and concepts can be applied to specific landscapes.

The K-cycle concept

Butler (1959) proposed the concept of the groundsurface to represent the development of a soil mantle. The chronologic order of groundsurfaces, set by their stratigraphic order, is K_1, K_2, K_3, K_4, etc, from the most recent (uppermost) to lower and older ones. A succession of buried soils indicates a recurrent cycle of stable and unstable phases of landscape evolution which Butler called a K-cycle. Each cycle will have an unstable phase (Ku) of erosion and deposition followed by a stable phase (Ks), accompanied by soil development. It is necessary to establish that each proposed groundsurface is an independent layer as a groundsurface must be shown by the development of a soil profile in it but, in some places, the layer may be too thin to contain the full A/B/C soil profile and must be traced laterally to a point where it is thick enough to recognize a full profile. The underlying deposits may have several horizon remnants of truncated soil profiles but no recognizable soil profile, and therefore, do not qualify as groundsurfaces. However, by tracing the deposit laterally, a complete soil profile might be recognized. This is just the problem of composite palaeosols stressed earlier and tests of independence must be continually applied.

Detailed traverses usually indicate that each groundsurface has a restricted range of soils in it. This relationship needs to be established before firm conclusions are reached; but once it is, the task of differentiating and delineating groundsurfaces is made very much easier. The difference between soils in groundsurfaces may not be great, but some feature is often different, and Butler argues that this feature should be given a high order of significance.

Hillslopes often show three zones of mantle activity: there is a sloughing zone of the steepest part where the soil mantle is completely stripped away at each erosive phase; below this and at slightly gentler angles is an alternating zone, where erosion has been less effective and the soils have not been completely truncated; the third zone, on the gentlest, lower slopes, is the accreting zone, where erosion between depositional phases has been nil and conformably superimposed groundsurfaces occur. These relationships are well illustrated in the situation described by Butler (1967) near Canberra, Australia (Figure 12.5). On the sloughing zones of the hillsides, K_2 groundsurface alone occurs. In the alternating zone, truncated remnants of K_3 and K_4 surfaces are found. The record of groundsurfaces is more complete on the toeslopes. The soil on K_1 surface is a dark prairie soil, soft and friable. Soils on groundsurface K_2 vary from red earths in well-drained sites to yellow earths and grey earths in poorly-drained sites. The red earth in the sloughing zone is often stony. This is clearly a detailed palaeocatena and again demonstrates that studies of palaeosols should be based on transects rather than individual sites. The catena of soils, in the K_3 groundsurface, has a red podzolic soil in well-drained sites and yellow podzolic soils, or solodic soils, in less well-drained areas. The soils possess a strongly contrasting pedal clay B horizon. The K_4 surface mostly occurs as a variably truncated buried layer. The catena is similar to K_3, but the B horizons are thicker with dense clay, and may include a calcareous gilgai in badly drained sites.

This sequence shows a complex sequence of buried, exhumed and relict palaeosols and requires very detailed and elaborate sampling and the application of precise stratigraphic principles to enable the correla-

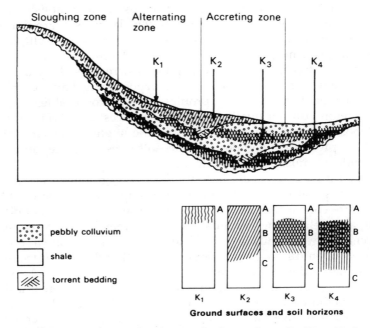

Figure 12.5 Sequence of groundsurfaces at Canberra, Australia (from Butler, 1967).

tions to be established. K_1 occurs as terraces, as small alluvial sheets and as segments in alluvial cones. K_2 occurs as hillside mantles on the lower two-thirds of slopes and as stony shallower soils higher up. K_3 occurs generally on the more extensive, gentler parts of the landscape. K_4 is usually buried. This arrangement, and the nature of the groundsurfaces, enable geomorphological conclusions to be made. The continuity, from hillslopes to terraces, indicated that terrace aggradation and erosion and deposition on hillsides were concurrent. It also seems that erosion and deposition were widespread over the whole landscape on several occasions. Initiation of each cycle was on the hillslopes and it is suggested that periods of instability were associated with more arid phases and the soils developed in intervening more humid phases.

The problems of establishing the relationships between relict and buried palaeosols and between distinct buried palaeosols were encountered by Walker (1962a) in New South Wales, Australia. Four soil layers can be identified. Soil layer M is a dark, loam to clay loam, seldom greater than 30 cm (12 in) thick and always associated with the steepest slopes. Soil layer N is a crumbly silt loam with a weak blocky structure. The profile is a grey-brown soil, characterized by slight A/B horizon differentiation and occurs also on steep slopes. Layer Wa is a dark grey-brown apedal clay loam and occurs on undulating terrain. Layer

Wb is a red plastic clay and generally occurs on gently sloping terrain. Where layers M, N, and Wb occur in association there is no pedogenetic relationship between them, only an ordered stratigraphy. An upper pedogenetic horizon may be removed by erosion but it will not occur without the lower horizons. This dependence was shown to exist between Wa and Wb and they are, therefore, the A and B horizons of one soil profile. Layer N is overlain by layer M but overlies Wb.

These soils represent a sequence of groundsurfaces in the manner suggested by Butler (1959). The uppermost stratigraphically and, therefore, the youngest is layer M and is designated K_1. Layer N represents the K_2 surface and the Wa/Wb layer is the oldest or K_3 groundsurface (Figure 12.6). On some gentle slopes K_3 (Wa/Wb) soils have persisted intact, through K_2 and K_1 cycles and are relict palaeosols. K_2 soils are relict on some slopes also. K_1 soils alone represent simple soils developed under the present environment. Because relict K_3 soils have not been affected by soil formation during K_2 and K_1 cycles and K_2 soils have not been altered by soil formation during the K_1 cycle, it appears that pedogenesis was stronger and/or lengthier in the K_3 cycle than in the K_2, and greater in the K_2 than in the K_1 cycle. This emphasizes the vexing questions discussed in a general way earlier in this chapter.

The soil layers just described are continuous with the stream deposits and are integral members of the geomorphological systems. There are also localities where the deposits are separated as terraces in a step-like

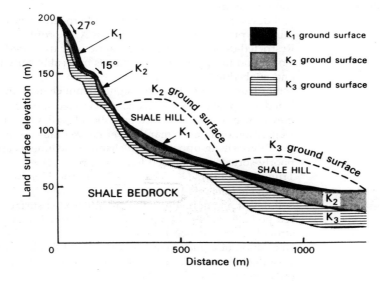

Figure 12.6 K-cycle soil layering on slopes in New South Wales, Australia (from Walker, 1962a).

succession. This implies that each layer is the result of a separate period of erosion and deposition. This relationship and the stratigraphical information allow a sequence of events to be established. Each unstable phase in the cycle involved the truncation and removal of soil material, with subsequent deposition on the lower slopes, as unstable phases waned. Erosional phases were accompanied by incision along the stream channels. A summary of the full sequence of events is given in Table 12.3. The transition between stable and unstable phases implies gross climatic changes. The differentiation of normal soil horizons in the layers could not have occurred without a well-vegetated surface and a climate suitable for the maintenance of the vegetation. Walker (1962a) also argues that the phases in instability, with sheetwash and gullying, could not have happened other than in a situation with sparse vegetation. A change from an equable humid climate to one of relatively dry conditions is indicated. This example not only illustrates the precise application of stratigraphic principles to soils but also demonstrates the geomorphological and environmental synthesis that is possible.

Table 12.3 Sequence of groundsurfaces and associated processes in eastern Australia (from Walker 1962a)

Cycle	Phase	Process
K_0	contemporary erosion and deposition	Slight channelling in the K_1 and K_2 deposits and deposition of stratified gravels, sands, and clays. No soil formation
K_1s		Soil development on depositional and eroded surfaces: minimal prairie soils formed with an A/C profile
K_1		Restricted deposition of a thin mantle over truncated slopes and in upper stream channels of weakly sorted and nonbedded sediment
K_1u		Minor truncation of K_2 soils; slight stream channelling
K_2s		Soil development on depositional and eroded surfaces; grey-brown soils formed with gradational A/B/C profiles
K_2		Deposition of an even mantle over eroded hillsides and thick deposits in upper stream channels; sediments are poorly-sorted and non-bedded
K_2u		Truncation of some K_3 soils to bedrock: considerable stream channelling
K_3s		Soil development on depositional and eroded surfaces; red podzolic and yellow podzolic soils formed with strongly differentiated profiles
K_3		Deposition of a thick mantle over eroded hillsides and very thick deposits in upper stream channel; sediments poorly-sorted and non-bedded
K_3u		Complete truncation of pre-K_3 soils to bedrock; very deep stream channelling

Northern Lake District, England

The Troutbeck buried soil, in the north-eastern Lake District, northern England, developed in pre-Devensian glacigenic sediments, allows a synthesis of landscape evolution to be proposed (Boardman, 1983, 1985b). The area is deeply dissected, with peaks reaching almost 1000 m and deep glacial troughs radiating from the central highland. The two major lithostratigraphic units are the glacigenic deposits of the Threlkeld and Thornsgill Formations. The Threlkeld Formation consists of till, sands and gravels and laminated beds. It is the local representative of the Late Devensian glaciation culminating about 18 000 BP. The Thornsgill Formation consists of a gravelly till and glacifluvial sands and gravels. The Troutbeck soil has developed in the Thornsgill Formation. The greatest depth of weathering is 15 m. In general the character of the weathered zone appears to represent the lower parts of a soil profile and indicate pedogenesis under temperate conditions on a stable, vegetated landsurface. Gley features suggest periods of prolonged seasonal saturation. Pedogenesis was halted by deposition of the overlying till and glacifluvial beds.

The degree of weathering of the buried soil suggests that soil development was aided by the incision of valleys into the land surface. The degree of weathering is also greater than that which has affected the overlying deposits during the Flandrian. A reconstruction of the landscape history is shown in Figure 12.7. The Troutbeck soil developed during phases 3–5, aided by localized valley incision. A later phase of incision, exposing the soil, occurred during the Loch Lomond Stadial (10 000–11 000 years BP). The Troutbeck soil is important because of a sparsity of information relating to pre-Devensian conditions during the Quaternary in northern England. Interglacial deposits, probably of Ipswichian age, have been described from Yorkshire (Gaunt *et al.* 1972) and the north-east (Beaumont *et al.*, 1969). There is also an interglacial pollen record from the north-west (Carter *et al.*, 1978) and evidence from North Yorkshire (Bullock, *et al.*, 1973) and north-east Scotland (Connell *et al.*, 1982). The survival of the Troutbeck soil is also important in that it implies low rates of glacial erosion in that part of the Lake District during the Late Devensian glaciation. Thus, it adds considerably to our knowledge of soils and Quaternary landscape evolution prior to the Late Devensian glaciation (Boardman, 1985b).

Landscape change in southern Iceland

Frequent volcanic eruptions, with the deposition of tephra, provide an excellent chronological sequence within which to examine landscape change in Iceland (Gerrard, 1985). Many of the tephra falls can be dated

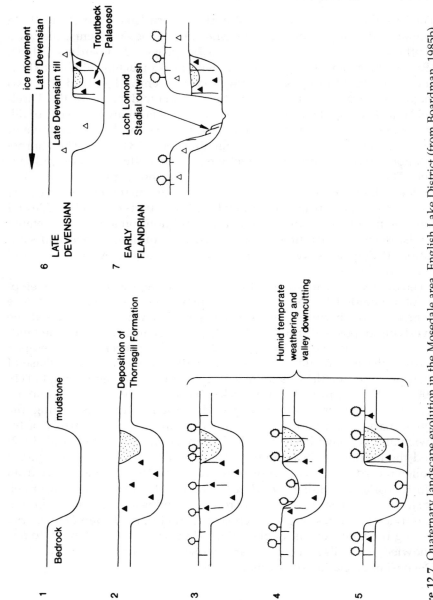

Figure 12.7 Quaternary landscape evolution in the Mosedale area, English Lake District (from Boardman, 1985b).

and they trap and preserve the soils, sediments and landforms on which they fall. A typical sequence of deposits from southern Iceland is shown in Figure 12.8. The Landnam tephra is so-called because it is thought to have occurred approximately at the time of Icelandic settlement (c AD 900). Thus, it forms an important marker horizon. Below the Landnam tephra the landscape has been evolving essentially without human influence, above it human influence becomes more important. This is

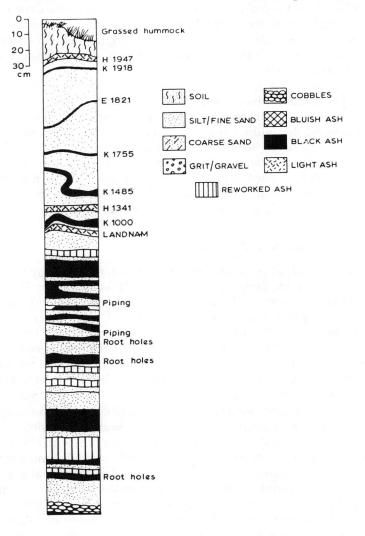

Figure 12.8 Soil section on the plateau area above Seljaland, southern Iceland (from Gerrard, 1985).

portrayed in the section. The increase in the amount of silt and fine sand above the Landnam ash is very clear. The rate of thickening of material below the Landnam ash shows an average annual rate of 0.14 mm yr^{-1}, whereas that above it shows an increase of 1.2 mm yr^{-1}. Also many of the tephra layers below the Landnam ash contain abundant plant roots indicating, at the very least, a basic soil and vegetation cover. This sequence is in accord with the general study of Thorarinsson (1962), which has shown rapidly increasing soil erosion in historic times. It has been estimated that the natural climax vegetation of birch forest has been reduced to about 1000 km^2 from an original estimated cover of 40 000 km^2. About half the area below 400 m is now practically devoid of soil (Thorarinsson, 1970).

Analysis of a number of soil sequences allows a regional picture of landscape change to be outlined (Gerrard, 1991a,b). Where exposures are frequent the evolution of a single slope or a small drainage basin can be established. Individual sequences also provide information on specific processes. In the sequence portrayed in Figure 12.8, buried thufur or frost hummocks, can be seen in the upper 1.5 m. However, there is no indication of buried thufur below the Katla 1500 tephra and it is possible to speculate that the influx of silt, a material susceptible to frost heave, has meant a greater development of thufur.

Otago Peninsula, New Zealand

The precise application of the principles of soil stratigraphy was found to be necessary by Leslie (1973) in his detailed study of Quaternary deposits and surfaces on the Otago Peninsula, New Zealand. Five types of soil and regolith had to be differentiated. These were *in situ* loess, loess colluvium, colluvial mixtures of loess and weathered volcanic rocks, solifluction deposits consisting of loess and weathered rock and saprolite derived from the volcanic rocks. The landsurfaces and the soils developed on them are not contemporaneous, but the establishment of stratigraphic units, based on the lithologies in the layered regolith, allowed Leslie to reconstruct the processes and climates operating in the past (Figure 12.9). Palaeosol morphology indicated that climate during the interstadial period was comparable to the present with *in situ* chemical weathering and soil development. The mid-stadial climate was cold and wet in the summer months. Erosion under periglacial conditions prevailed with cryoplanation on the summit slopes and solifluction on the midslopes. There was also some loess accumulation. The late stadial climate was warmer and moist. Loess deposition and solifluction ceased, creep and mass movement became significant and much of the earlier material was redeposited as colluvium. Radiocarbon dates and pollen from postglacial peats indicated a warm climate with forest species. This

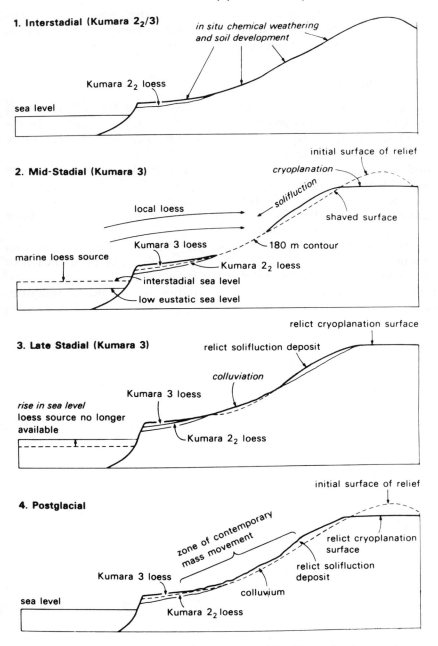

Figure 12.9 Sequence of geomorphic events and landform development on the Otago Peninsula, New Zealand (from Leslie, 1973, reproduced from *New Zealand Journal of Geology and Geophysics*).

forest cover persisted until about 1850 when the phase of European settlement resulted in forest clearance and caused mass movement. Landsliding is now a common feature of the area.

CONCLUSIONS

The identification and analysis of palaeosols and the application of stratigraphic principles enable environmental reconstruction to be established. The stratigraphic applications of soils are helped by a variety of relative and absolute dating techniques. But many of the techniques used are far from straightforward and considerable care has to be exercised. Also, the variability of the materials involved, including palaeosols, necessitates elaborate field sampling and the construction of detailed transects. It is clear that in many studies this has not been achieved. But, there are notable exceptions to this; the idea of K-cycles and the concept of the groundsurface provide a powerful research tool in studies of this nature. Many examples could have been chosen to illustrate the principles but all that has been attempted here has been a fairly detailed analysis of four specific studies that embody the majority of principles involved.

References

Aandahl, A.R. (1948) The characterisation of slope positions and their influence on the total nitrogen content of a few virgin soils of western Iowa. *Proceedings of the Soil Science Society of America*, **13**, 449–54.

Acton, D.F. (1965) The relationship of pattern and gradient of slopes to soil type. *Canadian Journal of Soil Science*, **45**, 96–101.

Acton, D.F. and Fehrenbacher, J.B. (1976) Mineralogy and topography of glacial tills and their effect on soil formation in Saskatchewan, in *Glacial till: an inter-disciplinary study* (ed R.T. Legget), Royal Society of Canada, Special Publication No 12.

Adams, G.F. (ed) (1975) *Planation surfaces: peneplains, pediplains and etchplains*, Dowden, Hutchinson and Ross, Stroudsberg, Pennsylvania.

Adams, W.A. and Raza, M.A. (1978) The significance of truncation in the evolution of slope soils in Mid-Wales. *Journal of Soil Science*, **29**, 243–57.

Afanasiev, J.N. (1927a) The Classification Problem in Russian Soil Science, in *Russian Pedological Investigations*, **5**, Academy of Sciences of the USSR, Lenigrad.

Afanasiev, J.N. (1927b) Soil classification problems in Russia. *Transactions of the 1st International Congress of Soil Science*, **4**, 498–501.

Agarwal, R.R., Mehotra, C.L. and Gupta, R.N. (1957) Development and morphology of Vindhyan soils: I. catenary relationship existing among the soils of the upper Vindhyan plateau, Uttar Pradesh. *Indian Journal of Agricultural Science*, **27**, 395–411.

Ahman, R. (1976) The structure and morphology of minerogenic palsas in northern Norway. *Biuletyn Peryglacjalny*, **26**, 25–31.

Ahn, P.M. (1970) *West African Agriculture, Vol. I: West African Soils*, 3rd edn, Oxford University Press, Oxford.

Åkerman, J. (1987) Periglacial forms of Svalbard: a review, in *Periglacial processes and landforms in Britain and Ireland*, (ed J. Boardman), Cambridge University Press, Cambridge, pp. 9–25.

Aleshinskaya, Z.V., Bondarev, L.G. and Gorbonov, A.P. (1972) Periglacial phenomena and some palaeo-geographical problems of Central Tien Shan. *Biuletyn Peryglacjalny*, **21**, 5–14.

Alexander, C.S. and Prior, J.C. (1971) Holocene sedimentation rates in overbank deposits in the Black Bottom of the Lower Ohio River, Southern Illinois. *American Journal of Science*, **270**, 361–72.

Alexander, E.B. (1985) Rates of soil formation from bedrock or consolidated sediments. *Physical Geography*, **6**, 25–42.

Alexander, E.B. (1988) Rates of soil formation: implications for soil-loss tolerance. *Soil Science*, **145**, 37–45.

Alexander, M.J. (1970) A study of some soils in the Austre Okstindsbredal area, in *Okstinden research project preliminary report, 1968*, (ed P. Worsley), Department of Geography, Reading University, pp. 25–31.

American Commission on Stratigraphic Nomenclature (1961) Code of stratigraphic nomenclature. *Bulletin of the Association of American Petroleum Geologists*, **45**, 645–65.

Anderson, H.W. (1954) Suspended sediment discharge as related to streamflow, topography, soil and land-use. *Transactions of the American Geophysical Union*, **35**, 268–81.

Anderson, K.E. and Furley, P.A. (1975) An assessment of the relationship between the surface properties of chalk soils and slope form using principal components analysis. *Journal of Soil Science*, **26**, 130–43.

Anderson, M.G. and Burt, T.P (1978a) Experimental investigations concerning the topographic control of soil water movement on hillslopes. *Zeitschrift für Geomorphologie, N.F. Supplementband*, **29**, 52–63.

Anderson, M.G. and Burt, T.P. (1978b) The role of topography in controlling throughflow generation. *Earth Surface Processes and Landforms*, **3**, 331–44.

Arnett, R.R. (1971) Slope form and geomorphological process: an Australian example. *Institute of British Geographers Special Publication*, **3**, 81–92.

Arnett, R.R. and Conacher, A.J. (1973) Drainage basin expansion and the nine unit landsurface model. *Australian Geographer*, **12**, 237–49.

Arnold, R.W. (1968) Pedological significance of lithologic discontinuities. *Transactions of the 9th International Congress of Soil Science*, **4**, 595–603.

Askew, G.P., Moffatt, D.J., Montgomery, R.E. and Searl, P.L. (1970) Soil landscapes in north eastern Mato Grosso. *Geographical Journal*, **136**, 211–27.

Augustinius, P.G.E.F. and Slager, S. (1971) Soil formation in swamp soils of the coastal fringe of Surinam. *Geoderma*, **6**, 203–11.

Avery, B.W. (1958) A sequence of beechwood soils on the Chiltern Hills, England. *Journal of Soil Science*, **9**, 210–24.

Baker, V.R., Kochel, R.C., Patton, P.C. and Pickup, G. (1983) Palaeohydrologic analysis of Holocene flood slackwater sediments. *International Association of Sedimentologists, Special Publication*, **6**, 229–39.

Bal, L. (1973) *Micromorphological analysis of soils*, Soil Survey Institute, Wageningen, The Netherlands.

Balme, O.E. (1953) Edaphic and vegetational zoning on the Carboniferous Limestone of the Derbyshire dales. *Journal of Ecology*, **41**, 331–44.

Bardossy, G. (1981) *Karst bauxites*, Akademiai Kiado, Budapest.

Barratt, B.C. (1962) Soil organic regime of coastal sand dunes. *Nature*, **196**, 835–7.

Basher, L.R., Tonkin, P.J. and Daly, G.T. (1985) Pedogenesis, erosion and revegetation in a mountainous high rainfall area – Cropp River, central Westland, in *Proceedings of the Soil Dynamics and Land Use Seminar, Blenheim, May 1985*, New Zealand Society of Soil Science and New Zealand Soil Conservationists Association, Blenheim, 49–64.

Basher, L.R., Tonkin, P.J. and McSaveney, M.J. (1988) Geomorphic history of a rapidly uplifting area on a compressional plate boundary, Cropp River, New Zealand. *Zeitschrift für Geomorphlogie, Supplementband*, **69**, 117–31.

Bateman, R.M. and Catt, J.A. (1985) Modification of heavy mineral assemblages in English coversands by acid pedochemical weathering. *Catena*, **12**, 1–21.

Beaumont, P., Turner, J. and Ward, P.F. (1969) An Ipswichian peat raft in glacial till at Hutton Henry, Co. Durham. *New Phytologist*, **68**, 797–805.

Beckett, P. (1968) Soil formation and slope development: I A review of W. Penck's aufbereitung concept. *Zeitschrift für Geomorphologie*, **12**, 1–24.

Benedict, J.B. (1969) Microfabric of patterned ground. *Arctic and Alpine Research*, **1**, 45–8.

Bennema, J. (1953) Die outkalking tijdens de opslibbing bij Nederlandse alluviale gronden. *Boor en Spade*, **6**, 30–40.

Berry, M.E. (1983) Morphological characteristics of soil catenas on Pinedale and Bull Lake moraines, Bear Valley, Idaho. Abstract *American Geomorphological Field Group Field Trip Guidebook*, p. 243.

Berry, M.E. (1984) Morphological and chemical characteristics of soil catenas on Pinedale and Bull Lake moraine slopes, Bear Valley (Salmon River Mountains), Idaho, unpublished M.S. thesis, University of Colorado, Boulder.

Bettenay, E. and Hingston, F.J. (1964) Development and distribution of soils in the Merredin area, Western Australia. *Australian Journal of Soil Research*, **2**, 173–86.

Bilzi, A.F. and Ciolkosz, E.J. (1977) A field morphology rating scale for evaluating pedological development. *Soil Science*, **124**, 45–8.

Birkeland, P.W. (1967) Correlation of soils of stratigraphic importance in western Nevada and California and their relative rates of profile development, in *Quaternary soils*, (eds R.R. Morrison and H.E. Wright), Desert Research Institute, University of Nevada, Reno, **9**, 71–91.

Birkeland, P.W. (1974) *Pedology, weathering and geomorphological research*, Oxford University Press, New York.

Birkeland, P.W. (1984) *Soils and geomorphology*, Oxford University Press, New York.

Birkeland, P.W. (1985) Quaternary soils of the western United States, in *Soils and Quaternary Landscape Evolution*, (ed J. Boardman), Wiley, Chichester, pp. 303–24.

Birkeland, P.W. (1990) Soil-geomorphic analysis and chronosequences – a selective overview. *Geomorphology*, **3**, 207–24.

Birot, P. (1960) *Le cycle d'érosion sous les différents climats*, Universidad do Brasil, Rio de Janeiro.

Birrell, K.S. and Pullar, W.A. (1973) Weathering of paleosols in Holocene and late Pleistocene tephras in central North Island, New Zealand. *New Zealand Journal of Geology and Geophysics*, **16**, 687–702.

Bishop, P.M., Mitchell, P.B. and Paton, T.R. (1980) The formation of duplex soils on hillslopes in the Sydney Basin, Australia. *Geoderma*, **23**, 175–89.

Biswas, T.D. and Gawande, S.P. (1962) Studies in genesis of catenary soils on sedimentary formation in Chatishgarh basin of Madhya Pradesh. I. Morphology and mechanical composition. *Journal of the Indian Society of Soil Science*, **10**, 223–34.

Bjorkhem, V. and Jongerius, A. (1974) Micromorphological observations in some podzolised soils from central Sweden, in *Soil microscopy*, (ed G.K. Rutherford), Limestone Press, Kingston, pp. 320–32.

Blake, D.H. and Ollier, C.D. (1971) Alluvial plains of the Fly River, Papua. *Zeitschrift für Geomorphologie*, **12**, 1–17.

Blokhuis, W.A., Slager, S. and Van Schagen, R.H. (1970) Plasmic fabrics of two Sudan Vertisols. *Geoderma*, **4**, 127–37.

Bloomfield, C. (1973) Some chemical properties of hydromorphic soils, in *Pseudogley and gley*, (eds E. Schlichtung and U. Schwertmann), Verlag Chemie, Weinheim, pp. 7–14.

Boardman, J. (1983) The role of micromorphological analysis in an investigation of the Troutbeck Paleosol, Cumbria, England, in *Soil Micromorphology*, (eds P. Bullock and C.P. Murphy), AB Academic Publishers, Berkhamsted, pp. 281–8.

Boardman, J. (ed) (1985a) *Soils and Quaternary Landscape Evolution*, Wiley, Chichester.

Boardman, J. (1985b) The Troutbeck Paleosol, Cumbria, England, in *Soils and Quaternary Landscape Evolution*, (ed J. Boardman), Wiley, Chichester, pp. 231–60.

Boardman, J. (ed) (1987) *Periglacial processes and landforms in Britain and Ireland*, Cambridge University Press, Cambridge.

Bockheim, J.G. (1980) Solution and use of chronofunctions in studying soil development. *Geoderma*, **24**, 71–85.

Bolyshev, N.N. (1964) Role of algae in soil formation. *Soviet Soil Science*, 1964, 630–5.

Bos, R.H.G. and Sevink, J. (1975) Introduction of gradational and pedomorphic features in descriptions of soils. *Journal of Soil Science*, **26**, 223–33.

Boulton, G.S. and Dent, D.L. (1974) The nature and rates of post-depositional changes in recently deposited till from south-east Iceland. *Geografiska Annaler*, **56a**, 121–34.

Bowler, J.M. and Polach, H.A. (1971) Radiocarbon analyses of soil carbonates: an evaluation from paleosols in southeastern Australia, in *Paleopedology*, (ed D.H. Yaalon), Israel Universities Press, Tel Aviv.

Braker, W.L. (1981) *Soil survey of Centre County, Pennsylvania*. USDA Soil Conservation Service, Washington.

Brammer, H. (1964) An outline of the geology and geomorphology of East Pakistan in relation to soil development. *Pakistan Journal of Soil Science*, **1**, 14–19.

Brammer, H. (1966) FAO/UNSF soil survey project of Pakistan progress of work in East Pakistan, 1961–1965. *Pakistan Journal of Soil Science*, **2**, 39–40.

Brammer, H. (1971) Coatings in seasonally-flooded soils. *Geoderma*, **5**, 5–16.

Brewer, R. (1968) Clay illuviation as a factor in particle size differentiation in the soil profile. *Transactions of the 9th International Congress of Soil Science*, **4**, 489–99.

Brewer, R. (1972) Use of macro- and micromorphological data in soil stratigraphy to elucidate surficial geology and soil genesis. *Journal of the Geological Society of Australia*, **19**, 331–44.

Brewer, R. and Pawluk, S. (1975) Investigations of some soils developed in hummocks of the Canadian sub-Arctic and southern Arctic regions. I. Morphology and micromorphology. *Canadian Journal of Soil Science*, **55**, 301–19.

Brewer, R. and Sleeman, J.R. (1960) Soil structure and fabric: their definition and description. *Journal of Soil Science*, **11**, 172–85.

Bricheteau, L. (1954) An example of sequence in red Mediterranean soils. *Bulletin Association Francaise Etude Sol*, **56**, 139–48.

Brinkman, R. (1970) Ferrolysis, a hydromorphic soil forming process. *Geoderma*, **3**, 199–206.

Brown, C.N. (1956) The origin of caliche on the north-east Llano Estacado, Texas. *Journal of Geology*, **64**, 1–15.

Brown, I.C. and Drossdoff, M. (1940) Chemical and physical properties of soils and their colloids developed from granitic materials in the Mojave Desert. *Journal of Agricultural Research*, **61**, 338–44.

Brown, I.C. and Thorp, J. (1942) *Morphology and composition of some soils of the Miami family and the Miami catena*. U.S. Department of Agriculture Technical Bulletin, 834.

Brown, J. (1966) Soils of the Okpilak River Region, Alaska. *Cold Regions Research and Engineering Laboratory Report*, **188**, US Army, Hanover, New Hampshire.

Brown, J. (1967) Tundra soils formed over ice wedges, northern Alaska. *Proceedings of the Soil Science Society of America*, **31**, 686–91.

Brown, J. and Tedrow, J.C.F. (1964) Soils of the northern Brooks Range, Alaska, 4: Well-drained soils of the glaciated valleys. *Soil Science*, **97**, 187–95.

Brown, R.J.E. (1969) Factors influencing discontinuous permafrost in Canada, in *The periglacial environment* (ed T.L. Pewe), McGill-Queens University Press, Montreal, pp. 11–53.

Brown, R.J.E. (1973) Influence of climate and terrain factors on ground temperatures at three locations in the permafrost region of Canada, in *Permafrost, North American Contribution, 2nd International Permafrost Conference, Yakutsk, USSR*, National Academy Science Publication 2115, pp. 27–34.

Brown, R.J.E. and Pewe, T.L. (1973) Distribution of permafrost in North America and its relationship to the environment; A review 1963–1973, in *Permafrost, North American Contribution, 2nd International Permafrost Conference, Yakutsk, USSR*, National Academy Science Publication 2115, pp. 71–100.

Brown, R.J.E. and Williams, G.P. (1972) *The freezing of peatlands*, Division of Building Research, National Research Council of Canada, Ottowa, Technical Paper 381.

Brown, T.M. and Kraus, M.J. (1981) Lower Eocene alluvial paleosols (Willwood Formation, northwest Wyoming, U.S.A.) and their significance for paleoecology, paleoclimatology and basin analysis. *Palaeogeography, Palaeoclimatology and Palaeoecology*, **34**, 1–30.

Bruce, J.G. (1973) Loessial deposits in southern South Island, with a definition of Stewarts Claim Formation. *New Zealand Journal of Geology and Geophysics*, **16**, 533–48.

Bruin, P. and Ten Have, J. (1935) Hat bepalen van magnesium carbonaat naast calcium carbonat in grond. *Chem Weekblad*, **32**, 375–8.

Brummer, G. (1973) Redoxreaktionen als merkmalspragende Prozesse hydromorpher Boden, in *Pseudogley and gley*, (eds E. Schlichtung and U. Schwertmann) Verlag Chemie, Weinheim, pp. 17–29.

Bryan, K. (1922) *Erosion and sedimentation in the Papago Country, Arizona*, US Geological Survey, Bulletin 730.

Bryant, R.H. and Carpenter, C.P. (1987) Ramparted ground ice depressions in Britain and Ireland, in *Periglacial processes and landforms in Britain and Ireland*, (ed J. Boardman) Cambridge University Press, Cambridge, pp. 183–90.

Bryant, R.H., Carpenter, C.P. and Ridge, T.S. (1985) Pingo scars and related features in the Whicham Valley, Cumbria, in *Field Guide to the periglacial landforms of Northern England* (ed J. Boardman), Quaternary Research Association, Cambridge, pp. 47–53.

Budel, J. (1957) Double surfaces of levelling in the humid tropics. *Zeitschrift für Geomorphologie*, **1**, 223–5.

Buffington, L.C. and Herbel, C.H. (1965) Vegetation changes on a semidesert grassland range. *Ecological Monographs*, **35**, 139–64.

Bull, W.B. (1962) Relation of textural (cm) patterns to depositional environment of alluvial-fan deposits. *Journal of Sedimentary Petrology*, **32**, 211–16.

Bull, W.B. (1968) Alluvial fan, in *The encyclopedia of geomorphology* (ed R. Fairbridge), Reinhold, New York, pp. 7–10.

Bull, W.B. (1972) Recognition of alluvial-fan deposits in the stratigraphic record, in *Recognition of ancient sedimentary environments*, Society of Economic Paleontologists and Mineralogists Special Publication, **16**, 63–83.

Bull, W.B. (1979) Threshold of critical power in streams. *Bulletin of the Geological Society of America*, **90**, 453–64.

Bull, W.B. (1990) Stream-terrace genesis: implications for soil development. *Geomorphology*, **3**, 351–67.

Bullock, P. (1971) The soils of the Malham Tarn area. *Field Studies*, **3**, 381–408.

Bullock, P. (1985) The role of micromorphology in the study of Quaternary soil processes, in *Soils and Quaternary landscape evolution*, (ed J. Boardman) Wiley, Chichester, pp. 45–68.

Bullock, P. and Murphy, C.P. (1979) Evolution of a paleo-argillic brown earth (Paleudalf) from Oxfordshire, England. *Geoderma*, **22**, 225–52.

Bullock, P., Carroll, D.M. and Jarvis, R.A. (1973) Paleosol features in Northern England. *Nature, Physical Sciences*, **242**, 53–4.

Bunting, B.T. (1965) *The geography of soil*, Hutchinson, London.

Bunting, B.T. (1977) The occurrence of vesicular structures in arctic and subarctic soils. *Zeitschrift für Geomorphologie*, **21**, 87–95.

Bunting, B.T. (1983) High Arctic soils through the microscope: prospect and retrospect. *Annals of the Association of American Geographers*, **73**, 609–16.

Bunting, B.T. and Fedoroff, N. (1974) Micromorphological aspects of soil development in the Canadian High Arctic, in *Soil microscopy* (ed G.K. Rutherford) Limestone Press, Kingston, pp. 350–65.

Buol, S.W., Hole, F.D. and McCracken, R.J. (1973) *Soil genesis and classification*, Iowa State Press, Ames, Iowa.

Buringh, P. and Edelman, C.H. (1955) Some remarks about the soils of the alluvial plain of Iraq, south of Baghdad. *Netherlands Journal of Agricultural Science*, **3**, 40–49.

Burke, R.M. and Birkeland, P.W. (1983) Toposequences and soil development as a relative dating tool on a two-fold chronosequence of eastern Sierra morainal slopes, California. *Abstracts with Programs, Geological Society of America*, **15**, 327.

Burns, S.F. and Tonkin, P.J. (1982) Soil-geomorphic models and the spatial distribution and development of alpine soils, in *Space and time in geomorphology, Binghamton Symposium in geomorphology*, 12, Allen and Unwin, London, pp. 25–43.

Burrough, P.A. (1981) The fractal dimensions of landscape and other data. *Nature*, **294**, 240–42.

Burrough, P.A. (1983) Multiscale sources of spatial variation in soil. I. The application of fractal concepts to nested levels of soil variation. II. A non-Brownian fractal model and its application in soil survey. *Journal of Soil Science*, **34**, 577–620.

Burton, R.G.O. (1976) Possible thermokarst features in Cambridgeshire. *East Midland Geographer*, **6**, 230–40.

Burton, R.G.O. (1987) The role of thermokarst in landscape development in eastern England, in *Periglacial processes and landforms in Britain and Ireland* (ed J. Boardman), Cambridge University Press, Cambridge, pp. 203–8.

Bushnell, T.M. (1943) Some aspects of the soil catena concept. *Proceedings of the Soil Science Society of America*, **7**, 466–76.

Bushnell. T.M. (1945) The 'catena-drainage profile' key-form as a frame of reference in soil classification. *Proceedings of the Soil Science Society of America*, **9**, 219–22.

Butler, B.E. (1950) A theory of prior streams as a causal factor of soil occurrence in the Riverine Plain of south-eastern Australia. *Australian Journal of Agricultural Research*, **1**, 231–52.

Butler, B.E. (1956) Parna, an aeolian clay. *Australian Journal of Science*, **18**, 145–51.

Butler, B.E. (1958) *Depositional systems of the Riverine Plain in relation to soils*, CSIRO Australia Soil Publication 10, Canberra, Australia.

Butler, B.E. (1959) *Periodic phenomena in landscapes as a basis for soil studies*, CSIRO Australia Soil Publication 14, Canberra, Australia.

Butler, B.E. (1967) Soil periodicity in relation to landform development in south-eastern Australia, in *Landform studies from Australia and New Guinea* (eds J. Jennings and J.A. Mabbutt), Cambridge University Press, Cambridge, pp. 231–55.

Butler, B.E. (1982) A new system for soil studies. *Journal of Soil Science*, **33**, 581–95.

Butler, B.E. and Hutton, J.T. (1956) Parna in the riverine plain of south-eastern Australia and the soils thereon. *Australian Journal of Agricultural Research*, **7**, 536–53.

Buursink, J. (1971) *Soils of Central Sudan*, Scholans and Jens, Utrecht.

Cailleux, A. (1959) Études sur l'érosion et la sédimentation en Guyana. *Mem. Serv. Carte Geol. Fr.*, 1959, 49–73.

Caine, N. (1974) The geomorphic processes of the alpine environment, in *Arctic and Alpine Environments*, (eds J.D. Ives and R.G. Barry), Methuen, London, pp. 721–48.

Callow, W.J., Baker, M.J. and Pritchard, D.H. (1964) National Physical Laboratory radiocarbon measurements: II. *Radiocarbon*, **6**, 25–30.

Campbell, C.A., Paul, E.A., Rennie, D.A. and McCallum, K.J. (1967) Applicability of the carbon-dating method of analysis to soil humus studies. *Soil Science*, **2**, 217–24.

Campbell, I.B. (1973) Pattern of variation in steepland soils: variation on a single slope. *New Zealand Journal of Science*, **16**, 413–34.

Campbell, I.B. (1975) Pattern of variation in steepland soils: soil differences in complex topography. *New Zealand Journal of Science*, **18**, 53–66.

Carson, M.A. and Kirkby, M.J. (1972) *Hillslope form and process*, Cambridge University Press, Cambridge.

Carter, B.J. and Ciolkosz, E.J. (1986) Sorting and thickness of waste mantle material in a sandstone spur in central Pennsylvania. *Catena*, **13**, 241–56.

Carter, P.A., Johnson, G.A.L. and Turner, J. (1978) An interglacial deposit at Scandal Beck, N.W. England. *New Phytologist*, **81**, 785–90.

Caseldine, C.J. (1983) Pollen analysis and rates of pollen incorporation into a radiocarbon-dated palaeopodzolic soil at Haugabreen, southern Norway. *Boreas*, **12**, 233–46.

Catalano, L.R. (1972) Datos hidrologicos del desierto de Atacama. *Decion Minas y Geologica Publication*, **35**, 10–16.

Catt, J.A. (1979) Soils and Quaternary geology. *Journal of Soil Science*, **30**, 607–42.

Catt, J.A. (1986) *Soils and Quaternary Geology: A handbook for field scientists*, Clarendon Press, Oxford.

Catt, J.A. (1987) Effects of the Devensian cold stage on soil characteristics and distribution in eastern England, in *Periglacial processes and landforms in Britain and Ireland*, (ed J. Boardman), Cambridge University Press, Cambridge, pp. 145–52.

Catt, J.A. and Penny, L.F. (1966) The Pleistocene deposits of Holderness, East Yorkshire. *Proceedings of the Yorkshire Geological Society*, **35**, 375–420.

Chandler, R.J. (1972) Periglacial mudslides in Vestspitzbergen and their bearing on the origin of fossil 'solifluction' shears in low angled clay slopes. *Quarterly Journal of Engineering Geology*, **5**, 225–41.

Charter, C.F. (1949) The detailed reconnaissance soils survey of the cocoa country of the Gold Coast. *Proceedings of the Cocoa Conference, London, 1949.*

Charter, C.F. (1958) *Report on the environmental conditions prevailing in Block A Southern Province, Tanganyika*, Ghana Department of Agriculture, Occasional Paper I.

Chartres, C.J. (1980) A Quaternary soil sequence in the Kennet Valley, Central Southern England. *Geoderma*, **23**, 125–46.

Childs, C.W. (1973) Patterns of total elements concentrations in Quaternary loess columns. *Proceedings INQUA Congress IX*, Abstract, 61–2.

Childs, C.W., Parfitt, R.L. and Lee, R. (1983) Movement of aluminium as an inorganic complex in some podzolised soils, New Zealand. *Geoderma*, **29**, 139–55.

Chorley, R.J. and Kennedy, B.A. (1971) *Physical geography: a systems approach*, Prentice-Hall, London.

Churchill, R.R. (1982) Aspect-induced differences in hillslope processes. *Earth Surface Processes and Landforms*, **7**, 171–82.

Ciolkosz, E.J., Carter, B.J., Hoover, M.T. *et al.* (1990) Genesis of soils and landscapes in the Ridge and Valley Province of central Pennsylvania. *Geomorphology*, **3**, 245–61.

Ciolkosz, E.J., Cronce, R.C., Cunningham, R.L. *et al.*, (1986a) Geology and soils of Nittany Valley. *Pennsylvania State University Agronomy Series*, **88**, 52pp.

Ciolkosz, E.J., Cronce, R.C. and Sevon, W.D. (1986b) Periglacial features in Pennsylvania. *Pennsylvania State University Agronomy Series*, **92**, 15pp.

Ciolkosz, E.J., Waltman, W.J., Simpson, T.W. *et al.* (1989) Distribution and genesis of soils of the northeastern United States. *Geomorphology*, **2**, 285–302.

Clark, M.J. (ed) (1988) *Advances in Periglacial Geomorphology*, Wiley, Chichester.

Clarke, G.R. (1954) *Soils in the Oxford Region*, Oxford University Press, Oxford.

Clarke, G.R. (1957) *The study of soil in the field*, 4th edn, Oxford University Press, Oxford.

Clayden, B. (1964) *The soils of the Middle Teign Valley district of Devon*. Bulletin of the Soil Survey of Great Britain, Harpenden.

Clayden, B. (1971) *Soils of the Exeter District*, Memoir of the Soil Survey of Great Britain, Harpenden.

Clements, T. *et al.* (1957) *A study of desert surface conditions*. Headquarters Quartermaster Research and Development Command, Environmental Protection Research Division, Technical Report, EP 53, Washington DC, 110pp.

Cloudsley-Thompson, J.L. and Chadwick, M.J. (1965) *Life in Deserts*, Foulis, London.

Coen, G.M., Pawluk, S. and Odynsky, W. (1966) The origins of bands in sandy soils of the stony plain area. *Canadian Journal of Soil Science*, **46**, 245–54.

Coleman, J.D., Farrar, D.M. and Marsh, A.D. (1964) The moisture characteristics, composition and structural analysis of a red clay soil from Kenya. *Geotechnique*, **14**, 262–76.

Conacher, A.J. and Dalrymple, J.B. (1977) The nine unit landsurface model: an approach to pedogeomorphic research. *Geoderma*, **18**, 1–154.

Conacher, A.J. and Dalrymple, J.B. (1978) Identification, measurement and interpretation of some pedogeomorphic processes. *Zeitschrift für Geomorphologie, Supplementband*, **29**, 1–9.

Connell, E.R., Edwards, K.J. and Hall, A.M. (1982) Evidence for two pre-Flandrian paleosols in Buchan, north-east Scotland. *Nature*, **297**, 1–3.

Cooke, R.U. and Warren, A. (1973) *Geomorphology in deserts*, Batsford, London.

Cooke, R.U., Brunsden, D., Doornkamp, J.C. *et al.* (1982) *Urban geomorphology in drylands*, Oxford University Press, Oxford.

Coque, R. (1962) *La Tunisie Pre-Saharienne, Étude Geomorphologique*, Colin, Paris.

Corless, J.F. and Ruhe, R.V. (1955) The Iowan terrace and terrace soils of the Nishnabotna Valley in western Iowa. *Proceedings of the Iowan Academy of Science*, **62**, 345–60.

Corte, A.E. (1978) Rock glaciers as permafrost bodies with a debris cover as an active layer. A hydrological approach, Andes of Mendoza, Argentina. *Proceedings of the 3rd International Permafrost Conference, Edmonton, Canada*, National Research Council, Ottowa, 163–9.

Cotton, C.A. (1961) The theory of savanna planation. *Geography*, **46**, 89–101.

Cox, J.E. and Mead, C.B. (1963) Soil evidence relating to post-glacial climate on the Canterbury Plains. *Proceedings of the New Zealand Ecological Society*, **10**, 28–38.

Coxon, P. and O'Callaghan, P. (1987) The distribution and age of pingo remnants in Ireland, in *Periglacial processes and landforms in Britain and Ireland*, (ed J. Boardman), Cambridge University Press, Cambridge.

Crampton, C.B. and Taylor, J.A. (1967) Solifluction terraces in South Wales. *Biuletyn Peryglacjalny*, **16**, 15–36.

Crickmay, C.H. (1933) The later stages in the cycle of erosion. *Geological Magazine*, **70**, 337–47.

Crocker, R.L. (1946) The soil and vegetation of the Simpson Desert and its borders. *Transactions of the Royal Society of South Australia*, **70**, 235–58.

Crocker, R.L. (1952) Soil genesis and the pedogenic factors. *Biological Review*, **27**, 139–68.

Crocker, R.L. and Dickson, B.A. (1957) Soil development on the recessional moraines of the Herbert and Mendenhall glaciers, south-eastern Alaska. *Journal of Ecology*, **45**, 169–85.

Crocker, R.L. and Major, J. (1955) Soil development in relation to vegetation and surface age at Glacier Bay, Alaska. *Journal of Ecology*, **43**, 427–48.

Crompton, E. (1966) *The soils of the Preston District of Lancashire*. Memoir of the Soil Survey of Great Britain, Harpenden.

Culling, W.E.H. (1986) Highly erratic spatial variability of soil pH on Iping Common, West Sussex. *Catena*, **13**, 81–98.

Currey, D.T. (1977) The role of applied geomorphology in irrigation and groundwater studies, in *Applied Geomorphology* (ed. J.R. Hails), Elsevier, Amsterdam.

Curtis, C.D. (1976) Stability of minerals in surface weathering reactions. *Earth Surface Processes*, **1**, 63–70.

Curtis, L.F. (1971) *Soils of Exmoor Forest*, Soil Survey of Great Britain, Special Survey, 5, Harpenden.

Czudek, T. and Demek, J. (1970) Pleistocene cryopedimentation in Czechoslovakia. *Acta Geographica Lodziensia*, **24**, 101–8.

Czudek, T. and Demek, J. (1973) The valley cryopediments in Eastern Siberia. *Biuletyn Peryglacjalny*, **22**, 117–30.

Dalrymple, J.B., Blong, R.J. and Conacher, A.J. (1968) A hypothetical nine unit landsurface model. *Zeitschrift für Geomorphologie*, **12**, 60–76.

Dalsgaard, K., Baastrup, E. and Bunting, B.T. (1981) The influence of topography on the development of Alfisols on calcareous till in Denmark. *Catena*, **8**, 111–36.

Dan, J. and Yaalon, D.H. (1968) Pedomorphic forms and pedomorphic surfaces. *Transactions of the 9th International Congress of Soil Science, Adelaide*, **3**, 577–84.

Dan, J. and Yaalon, D.H. (1982) Automorphic saline soils in Israel. *Catena, Supplement*, **1**, 103–15.

Daniels, R.B. and Gamble, E.E. (1978) Relations between stratigraphy, geomorphology and soils in Coastal Plain areas of southeastern USA. *Geoderma*, **21**, 41–65.

Daniels, R.B., Gamble, E.E. and Cady, J.G. (1970) Some relations among Coastal Plain soils and geomorphic surfaces in North Carolina. *Proceedings of the Soil Science Society of America*, **34**, 648–53.

Daniels, R.B., Gamble, E.E. and Nelsom, L.A. (1971) Relations between soil morphology and water-table levels on a dissected North Carolina Coastal Plain surface. *Proceedings of the Soil Science Society of America*, **35**, 781–4.

Daniels, R.B., Gamble, E.E. and Wheeler, W.H. (1978) Ages of soil landscapes in the Coastal Plain of North Carolina. *Journal of the Soil Science Society of America*, **42**, 98–105.

Davis, W.M. (1899) The geographical cycle. *Geographical Journal*, **14A**, 481–503.

De Gans, W. (1981) *The Drentshce Aa Valley System. A study of Quaternary geology*, Vrije Universiteit te Amsterdam, 132pp.

Delcourt, H.R. and Delcourt, P.A. (1986) Late Quaternary vegetational change in the Central Atlantic States, in *The Quaternary of Virginia – A Symposium Volume*, (ed J.N. McDonald and S.O. Bird), Virginia Division of Mineral Resources Publication, 75, 23–35.

Delcourt, P.A. and Delcourt, H.R. (1983) Late-Quaternary vegetational dynamics and community stability reconsidered. *Quaternary Research*, **19**, 265–71.

Delvigne, J. (1965) *Pedogenese en zone tropicale. La formatin des mineraux secondaires en milieu ferrallitique*. Memoir, ORSTOM, **13**.

DeMumbrum, L.E. and Bruce, R.R. (1960) Mineralogy of three alluvial soils of the Mississippi River alluvial plain. *Soil Science*, **89**, 333–37.

De Smet, L.A.H. (1954) Enkele opmerkingen over kalkarme zeekleiafzettingen. *Boor en Spade*, **7**, 169–73.

De Swardt, A.M.J. (1964) Lateritisation and landscape development in parts of Equatorial Africa. *Zeitschrift für Geomorphologie*, **NS 8**, 313–33.

Diephuis, J.G.H.R. (1966) The Guiana Coast. *Tijdschr. Koninkl. Ned. Aardrijkskundig Genoot*, **83**, 145–52.

Dijkerman, J.C. (1974) Pedology as a science: the role of data, models and theories in the study of natural soil systems. *Geoderma*, **11**, 73–93.

Dimbleby, G.W. (1952) The historical status of moorland in northeast Yorkshire. *New Phytologist*, **56**, 12–28.

Dimbleby, G.W. (1961a) Soil pollen analysis. *Journal of Soil Science*, **12**, 1–11.

Dimbleby, G.W. (1961b) Transported material in the soil profile. *Journal of Soil Science*, **12**, 12–22.

Dimo, V.H. (1965) Formation of a humic-illuvial horizon in soils on permafrost. *Soviet Soil Science*, **9**, 1013–21.

Dixon, J.B. and Weed, S.B. (eds) (1977) *Minerals in soil environments*, Soil Science Society of America, Madison.

Doeglas, D.J. (1950) Die interpretatie van karrelgroote-analysen II: Karrelgroote-ouderzoek van Nederlandse strandzanden. *H. koninkl. Ned. Mijmbouwk. Genoot. Geol. Ser.*, **15**, 257–75.

Dokuchaev, V.V. (1879) Short historical description and critical analysis of the more important existing soil classifications. *Trav. Soc. Nat. (St. Petersburg)*, **10**, 64–7.

Dokuchaev, V.V. (1893) *The Russian Steppes and the study of the soil in Russia, its past and present*. J.W. Crawford, translation, Department of Agriculture, Ministry of Crown Domains for the World's Columbian Exposition, St. Petersburg, 61pp.

Dokuchaev, V.V. (1898) *The problem of the re-evaluation of the land in European and Asiatic Russia*, (in Russian), Moscow.

Doolittle, J.A. (1982) Characterising soil map units with the ground penetrating radar. *Soviet Soil Horizons*, **23**, 3–10.

Dormaar, J.F. (1967) Infrared spectra of humic acids from soils under grass or trees. *Geoderma*, **1**, 37–45.

Dormaar, J.F. (1973) A diagnostic technique to differentiate between buried Gleysolic and Chernozemic B horizons. *Boreas*, **2**, 13–15.

Dormaar, J.F. and Lutwick, L.E. (1969) Infrared spectra of humic acids and opal phytoliths as indicators of paleosols. *Canadian Journal of Soil Science*, **49**, 29–37.

Douglas, L.A. and Tedrow, J.C.F. (1960) Tundra soils of arctic Alaska. *Transactions of the 7th International Congress of Soil Science*, **4**, 291–304.

Dryden, L. and Dryden, C. (1946) Comparative rates of weathering of some common heavy minerals. *Journal of Sedimentary Petrology*, **16**, 91–6.

Duchaufour, P. (1982) *Pedology*. (trans. T.R. Paton), Allen and Unwin, London.

Dumanski, J. and St. Arnaud, R.J. (1966) A micromorphological study of eluviated horizons. *Canadian Journal of Soil Science*, **46**, 287–92.

Dunne, T. (1978) Field studies of hillslope flow processes, in *Hillslope Hydrology* (ed. M.J. Kirkby), J. Wiley and Sons, London, pp. 227–93.

Edelman, C.H. (1950) *Soils of the Netherlands*, North-Holland, Amsterdam.

Edelman, C.H. and Van der Voorde, Pk.J. (1963) Important characteristics of alluvial soils in the tropics. *Soil Science*, **95**, 258–63.

El-Attar, H. and Jackson, M.L. (1973) Montmorillonite soils developed in Nile River sediments. *Soil Science*, **116**, 191–201.

El-Gabaly, M. and Khadr, M. (1962) Clay mineral analysis of some Egyptian Desert and Nile alluvial soils. *Journal of Soil Science*, **13**, 333–42.

Ellis, J.H. (1938) *The soils of Manitoba*, Manitoba Economic Survey Board, Winnipeg.

Ellis, S. (1979) The identification of some Norwegian mountain soil types. *Norsk Geografisk Tidsskrift*, **33**, 205–11.

Ellis, S. (1980a) An investigation of weathering in some arctic-alpine soils on the northeast flank of Oksskolten north Norway. *Journal of Soil Science*, **32**, 371–85.

Ellis, S. (1980b) Physical and chemical characteristics of a podzolic soil formed in neoglacial till, Okstindan, northern Norway. *Arctic and Alpine Research*, **12**, 65–72.

Ellis, S. (1983) Micromorphological aspects of arctic-alpine pedogenesis in the Okstindan Mountains, Norway. *Catena*, **10**, 133–48.

Ellis, S. and Matthews, J.A. (1984) Pedogenic implications of a ^{14}C-dated paleopodzolic soil at Hangabreen, southern Norway. *Arctic and Alpine Research*, **16**, 77–91.

Ellison, W.D. (1944) Two devices for measuring soil erosion. *Agricultural Engineering*, **25**, 53–5.

England, C.B. and Holtan, H.N. (1969) Geomorphic grouping of soils in watershed engineering. *Journal of Hydrology*, **7**, 217–25.

English, C. (1977) An investigation into regolith thickness and its possible relationships with slope angle and slope position on the Upper Chalk of the South Downs, Sussex, unpublished B.Sc. dissertation, Department of Geography, University of Birmingham.

Eriksson, E. (1958) The chemical climate and saline soils in the arid zone. *Arid Zone Research*, **10**, 147–88.

Evans, C.J. (1978) Quantification and pedological processes, in *Quaternary soils* (ed W.C. Mahaney), Geo Abstracts, Norwich, pp. 361–78.

Faegri, K. and Iversen, J. (1974) *Textbook of pollen analysis*, Blackwell Scientific, Oxford.

FAO-UNESCO (1974) *FAO-UNESCO Soil map of the world, 1:5,000,000 Vol. 1.* Legend sheet and memoir, Paris.

Federova, N.N. and Yarilova, T. (1972) Micromorphology and genesis of pro-longed seasonally frozen soils in western Siberia. *Geoderma*, **7**, 1–13.

Fenwick, I.M. (1980) Palaeosols, in *Geomorphological Techniques* (ed A. Goudie), Allen and Unwin, London, pp. 342–5.

Fenwick, I.M. (1985) Paleosols: Problems of recognition and interpretation, in *Soils and Quaternary Landscape Evolution* (ed J. Boardman), Wiley, Chichester, pp. 3–21.

Findlay, D.C. (1965) *Soils of the Mendip District of Somerset*. Memoir of the Soil Survey of Great Britain, Harpenden.

Finkl, C.W. Jr. (1980) Stratigraphic principles and practices as related to soil mantles. *Catena*, **7**, 169–94.

Finkl, C.W. and Gilkes, R.J. (1976) Relationships between micromorphological soil features and known stratigraphic layers in Western Australia. *Geoderma*, **15**, 179–208.

Finney, H.R., Holowaychuk, N. and Heddleson, M.R. (1962) The influence of microclimate on the morphology of certain soils of the Allegheny Plateau of Ohio. *Proceedings of the Soil Science Society of America*, **26**, 287–92.

Firman, J.B. (1968) Soil distribution – a stratigraphic approach. *Transactions of the 9th International Congress of Soil Science*, **4**, 569–76.

Fishk, F.M., El-Attar, H.A., Hassan, M.N. *et al.* (1976) Mineralogical and chemical composition of the clay fraction of some Nile alluvial soils in Egypt. *Chemical Geology*, **17**, 295–306.

Fitze, P. (1981) Zur bodentwicklung auf morainen in den Alpen. *Bull. Bodenkundliche Gessellschaft Schueiz*, **5**, 29–34.

FitzPatrick, E.A. (1956) An indurated soil horizon formed by permafrost. *Journal of Soil Science*, **7**, 248–54.

FitzPatrick, E.A. (1980) *Soils: Their formation, classification and distribution*, Longman, London.

Follmer, L.R. (1978) The Sangamon Soil in its type area – a review, in *Quaternary soils* (ed W.C. Mahaney), Geo Abstracts, Norwich, pp. 125–65.

Fookes, P.G. and Knill, J.L. (1969) The application of engineering geology in the regional development of northern and central Iran. *Engineering Geology*, **3**, 81–120.

Foss, J.E. and Rust, R.H. (1962) Soil development in relation to loessial deposition in southeastern Minnesota. *Proceedings of the Soil Science Society of America*, **26**, 270–74.

Fowler, M. (1990) *The validity of the nine unit landsurface model to a steepland watershed in the Coast Mountains of southwestern British Columbia, Canada*. Unpublished BA dissertation, School of Geography, University of Birmingham.

Fox, C.A. and Protz, R. (1981) Definition of fabric distributions to characterise the rearrangement of soil particles in turbic cryosols. *Canadian Journal of Soil Science*, **61**, 29–34.

Franzmeier, D.P. and Whiteside, E.P. (1963) A chronosequence of podzols in Northern Michigan: I Physical and chemical properties. *Michigan State University Agriculture Experimental Station Quarterly Bulletin*, **46**, 21–36.

Franzmeier, D.P., Pedersen, E.J., Longwell, T.J. *et al* (1969) Properties of some soils in the Cumberland Plateau as related to slope aspect and position. *Proceedings of the Soil Science Society of America*, **33**, 755–61.

French, H.M. (1970) Soil temperatures in the active layer, Beaufort Plain. *Arctic*, **23**, 229–39.

French, H.M. (1971) Slope asymmetry of the Beaufort Plain, northwest Banks Island, N.W.T., Canada. *Canadian Journal of Earth Science*, **8**, 717–31.

French, H.M. (1976) *The periglacial environment*, Longman, London.

French, H.M. (1987) Periglacial processes and landforms in the Western Canadian Arctic, in *Periglacial Processes and Landforms in Britain and Ireland* (ed J. Boardman), Cambridge University Press, Cambridge, pp. 27–43.

Fridland, V.M. (1974) Structure of the soil mantle. *Geoderma*, **12**, 35–41.

Frye, J.C., Wilman, H.B. and Glass, H.D. (1960) *Gumbotil, accretion-gley and the weathering profile*, Illinois State Geological Survey Circular, 295.

Fuller, W.H., Cameron, R.E. and Raica, N. Jr. (1960) Fixation of nitrogen in desert soils by algae. *Transactions of the 7th International Congress of Soil Science, Madison*, **2**, 617–24.

Furbish, D.J. (1983) Use of a diffusion model of degradation and A-horizon soil properties to estimate river terrace age. Abstract *American Geomorphological Field Group Field Trip Guidebook*, 245–6.

Furley, P.A. (1968) Soil formation and slope development: 2 The relationship between soil formation and gradient angle in the Oxford area. *Zeitschrift für Geomorphologie*, **NF 12**, 25–42.

Furley, P.A. (1971) Relationships between slope form and soil properties developed over chalk parent materials, in *Slopes, form and process* (ed D. Brunsden), Institute of British Geographers Special Publication, 3, 141–64.

Gaunt, G.D., Coope, G.R., Osborne, P.J. *et al.* (1972) An interglacial deposit near Austerfield, southern Yorkshire. *Report Institute of Geological Sciences*, 72–4.

Gennadiyev, A.N. (1978) A study of soil formation by the chronosequence method as exemplified by the soils of the Elbrus region. *Soviet Soil Science*, **10**, 707–16.

Gerrard, A.J. (1978) Hillslope profile analysis. *Area*, **10**, 129–30.

Gerrard, A.J. (1982) Slope form and regolith characteristics in the basin of the River Cowsic, Central Dartmoor, Devon, unpublished Ph.D. thesis, University of London.

Gerrard, A.J. (1985) Soil erosion and landscape stability in southern Iceland: a tephrochronolgical approach, in *Geomorphology and Soils* (eds K.S. Richards, R.R. Arnett and S. Ellis), Allen and Unwin, London, pp. 78–95.

Gerrard, A.J. (1987) *Alluvial Soils*, Van Nostrand Reinhold, New York.

Gerrard, A.J. (1988a) Hillslope profile analysis: a comparison of different methods using slopes on Dartmoor, England, in *Geomorphology and Environment* (eds S. Singh and R.C. Tiwari), Allahabad Geographical Society, Allahabad, India, pp. 375–85.

Gerrard, A.J. (1988b) Periglacial modification of the Cox Tor – Staple Tors area of western Dartmoor, England. *Physical Geography*, **9**, 280–300.

Gerrard, A.J. (1988c) *Soil–slope relationships: A Dartmoor example.* Occasional Publication, School of Geography, The University of Birmingham, 26.

Gerrard, A.J. (1989a) The nature of slope materials on the Dartmoor granite. *Zeitschrift für Geomorphologie*, **NF 33**, 179–88.

Gerrard, A.J. (1989b) Partially infilled gully systems on Dartmoor. *Proceedings of the Usher Society*, **7**, 86–9.

Gerrard, A.J. (1990a) *Mountain Environments*, Belhaven Press, London.

Gerrard, A.J. (1990b) *Soil–landform relationships III. Examples of soil toposequences from the Chilterns and Cotswold Hills.* Occasional Publication, School of Geography, The University of Birmingham, 31.

Gerrard, A.J. (1990c) Soil variations on hillslopes in humid temperate climates. *Geomorphology*, **3**, 225–44.

Gerrard, A.J. (1991a) An assessment of some of the factors involved in recent landscape change in Iceland, in *Environmental change in Iceland: Past and present* (eds J.K. Maizels and C. Caseldine), Kluwer, Amsterdam, pp. 237–53.

Gerrard, A.J. (1991b) Landscape change in Berufjordur, Eastern Iceland in the last thousand years. *Indian Journal of Landscape Systems and Ecological Studies*, **13**, 1–9.

Gerrard, A.J. (1991c) The status of temperate hillslopes in the Holocene, *The Holocene*, **1**, 86–90.

Gerrard, A.J. (1992) The nature and geomorphological relationships of earth hummocks (thufur) in Iceland. *Zeitschrift für Geomorphologie, Supplementband*, **86**, 169–78.

Gerrard, A.J. (in prep.) *The climatic significance of earth hummocks.*

Gerrard, A.J. and Baker, S. (1990) *Soil–landform relationships. II. Variability of soils in the Fyn Valley, East Suffolk.* Occasional Publication, School of Geography, The University of Birmingham, 30.

Gerrard, A.J. and English, C. (1990) Soil–landform relationships. I. Soil thickness, slope angle and position on Chalk slopes of the South Downs, Sussex. Occasional Publication, School of Geography, The University of Birmingham, 29.

Gerrard, A.J. and Robinson, D.A. (1971) Variability in slope measurements. *Transactions of the Institute of British Geographers*, **54**, 49–54.

Gewaifel, I.M. and Younis, M.G. (1978) A comparative pedogenic study of some Nile alluvial soils in Egypt and Sudan. *Egyptian Journal of Soil Science*, **18**, 205–15.

Gewaifel, I.M., Hassan, M.N. and El-Gabaly, M.M. (1970) Clay mineralogy of some soil profiles from Western Desert (UAR). *Alexandria Journal of Agricultural Research*, **2**, 269–76.

Gewaifel, I.M., Younis, M.G. and El Zahaby, E.L. (1981) Heavy minerals study of some Nile alluvial soils in Egypt and Sudan. *Egyptian Journal of Soil Science*, **21**, 1–7.

Gibbs, H.S. (1980) *New Zealand Soils: An Introduction*, Oxford University Press, Wellington, New Zealand.

Gigon, A. (1983) Typology and principles of ecological stability and instability. *Mountain Research and Development*, **3**, 95–102.

Gigout, M. (1960) Sur la génése des croutes calcaires Pleistocenes en Afrique du Nord. *Société Geologique de France, CRS*, 8–10.

Gilbert, C.J. (1933) The evolution of Romney Marsh. *Arch. Cant.*, **45**, 246–72.

Gile, L.H. (1967) Soils of an ancient basin floor near Las Cruces, New Mexico. *Proceedings of the Soil Science Society of America*, **34**, 465–72.

Gile, L.H. and Grossman, R.B. (1968) Morphology of the argillic horizon in desert soils of southern New Mexico. *Soil Science*, **106**, 6–15.

Gile, L.H. and Grossman, R.B. (1979) *The Desert Project soil monograph*, United States Department of Agriculture, Soil Conservation Service, Washington DC.

Gile, L.H. and Hawley, J.W. (1966) Periodic sedimentation and soil formation on an alluvial fan piedmont in southern New Mexico. *Proceedings of the Soil Science Society of America*, **30**, 261–8.

Gile, L.H. and Hawley, J.W. (1968) Age and comparative development of desert soils at the Gardner Spring radiocarbon site, New Mexico. *Proceedings of the Soil Science Society of America*, **32**, 709–16.

Gile, L.H., Hawley, J.W. and Grossman, R.B. (1981) Soils and geomorphology in the Basin and Range area of southern New Mexico – Guidebook to the Desert Project. *New Mexico Bureau of Mines and Minerals Resources, Memoir*, 39.

Gile, L.H., Peterson, F.F. and Grossman, R.B. (1965) The K-horizon – a master horizon of $CaCO_3$ accumulation. *Soil Science*, **99**, 74–82.

Gile, L.H., Peterson, F.F. and Grossman, R.B. (1966) Morphological and genetic sequences of carbonate accumulation in desert soils. *Soil Science*, **10**, 347–60.

Gillham, M.E. (1957) Coastal vegetation of Mull and Iona in relation to salinity and soil reaction. *Journal of Ecology*, **45**, 157–75.

Ginsburg, R.N., Isham, L.B., Bein, S.J. *et al.* (1954) *Laminated algal sediments of south Florida and their recognition in the fossil record*, Marine Laboratory, University of Miami, Coral Gables, Florida, report 54–21.

Glazovskaya, M.A. (1968) Geochemical landscapes and types of geochemical soil sequences. *Transactions of the 9th International Congress of Soil Science*, **4**, 303–12.

Glennie, K.W. (1970) *Desert sedimentary environments*, Elsevier, Amsterdam.

Glentworth, R. (1954) *The soils of the country round Banff, Huntly and Turrif*, Memoir of the Soil Survey of Great Britain (Scotland), Harpenden.

Glentworth, R. and H.G. Dion (1949) The association or hydrologic sequence in certain soils of the podzolic zone of north-east Scotland. *Journal of Soil Science*, **1**, 35–49.

Goh, K.M. (1972) Amino acid levels as indicators of paleosols in New Zealand soil profiles. *Geoderma*, **7**, 33–47.

Goldich, S.S. (1938) A study of rock weathering. *Journal of Geology*, **46**, 17–58.

Goosen, D. (1972) *Physiography and soils of the Llanos Orientales of Colombia*, ITC, Enschede.

Goss, D.W. and Allen, B.L. (1968) A genetic study of two soils developed on granite in Llanos County, Texas. *Proceedings of the Soil Science Society of America*, **32**, 409–13.

Goudie, A.S. (1971) Calcrete as a component of semi-arid landscapes, unpublished PhD. thesis, University of Cambridge.

Goudie, A.S. (1973) *Duricrusts in tropical and sub-tropical landscapes*, Clarendon Press, Oxford.

Goudie, A.S. (1983) Calcrete, in *Chemical sediments and geomorphology* (eds A.S. Goudie and K. Pye), Academic Press, London, pp. 93–131.

Goudie, A.S. (1985) Duricrusts and landforms, in *Geomorphology and Soils* (eds K.S. Richards, R.R. Arnett and S. Ellis), Allen and Unwin, London, pp. 37–57.

Goudie, A.S. and Wilkinson, J. (1978) *The warm desert environment*, Cambridge University Press, Cambridge.

Grant, P.J. (1981) Major periods of erosion and sedimentation in the North Island, New Zealand since 13th century, in *Erosion and sediment transport in Pacific Rim Steepland*, International Association of Hydrological Sciences, **132**, 288–304.

Gray, A.J. and Bunce, R.G.H. (1972) The ecology of Morecambe Bay. VI. Soils and vegetation of the salt marshes: a multivariate approach. *Journal of Applied Ecology*, **9**, 221–34.

Green, P. (1974) Recognition of sedimentary characteristics in soils by size-shape analysis. *Geoderma*, **11**, 181–93.

Green, R.D. (1968) *Soils of Romney Marsh*. Bulletin of the Soil Survey of Great Britain, 4, Harpenden.

Green, R.D. and Askew, G.P. (1965) Observations on the biological development of macropores in soils of Romney Marsh. *Journal of Soil Science*, **16**, 342–9.

Greene, H. (1947) Soil formation and water movement in the tropics. *Soils and Fertilizer*, **10**, 253–6.

Gregory, K.J. and Walling, D. (1973) *Drainage basin form and process*, Edward Arnold, London.

Greig-Smith, P. and Chadwick, M.J. (1965) Data on pattern within plant communities. III. Acacia Capparis semi-desert scrub in the Sudan. *Journal of Ecology*, **53**, 465–74.

Griffiths, G.A. (1981) Some suspended sediment yields from South Island catchments, New Zealand. *Water Resources Bulletin*, **17**, 662–71.

Gruner, J.W. (1950) An attempt to arrange silicates in the order of reaction series at relatively low temperatures. *American Mineralogist*, **35**, 137–8.

Gupta, R.N. (1958) Development and morphology of Vindhyan soils. II. Nature of exchangeable bases in the soil sequence of the Upper Vindhyan plateau in Uttar Pradesh. *Indian Journal of Agricultural Science*, **28**, 491–8.

Gwynne, C.S. (1942) Swell and swale pattern of the Wisconsin Drift Plain of Iowa. *Journal of Geology*, **50**, 200–208.

Gwynne, C.S. and Simonson, R.W. (1942) Influence of low recessional moraines on soil type pattern of the Mankato Drift Plain in Iowa. *Soil Science*, **53**, 461–6.

Hack, J.T. (1960) Interpretation of erosional topography in humid climates. *American Journal of Science*, **258**, 80–97.

Hack, J.T. (1980) Dynamic equilibrium and landscape evolution, in *Theories of Landform Development* (eds W.N. McChorn and R.C. Flemal), Allen and Unwin, London, pp. 87–102.

Hack, J.T. and Goodlett, J.G. (1960) *Geomorphology and forest ecology of a mountain region in the Central Appalachians*, United States Geological Survey Professional Paper, no. 347.

Hall, B.R. and Folland, C.J. (1970) *Soils of Lancashire*. Bulletin of the Soil Survey of Great Britain, no. 5, Harpenden.

Hall, G.F. (1983) Pedology and geomorphology, in *Pedogenesis and soil taxonomy. I. Concepts and interactions* (eds L.P. Wilding, N.E. Smeck and G.F. Hall), Elsevier, Amsterdam, pp. 117–40.

Hallsworth, E.G. and Waring, H.D. (1964) Studies of pedogenesis in New South Wales. VIII. An alternative hypothesis for the formation of podzolised-solonetz of the Pilliga District. *Journal of Soil Science*, **15**, 158–77.

Hamdi, H. (1954) Mineralogical study of the alluvial suspended matter of the Nile. *Clay Minerals Bulletin*, **12**, 209–19.

Hamdi, H. (1959) Alteration in the clay fraction of Egyptian soils. *Z. Pflanzenernahr. Dung. Bodenkd.*, **84**, 204–11.

Hamdi, H. (1967) The mineralogy of the fine fraction of the alluvial soils of Egypt. *Journal of Soil Science of the United Arab Republic*, **7**, 15–21.

Hamdi, H. and Barrada, Y. (1960) Transformation of the clay fraction of the alluvial soils of Egypt. *Annals of Agricultural Science, Cairo*, **2**, 135–9.

Hamdi, H. and Iberg, R. (1954) Information on the alluvial clay of the Nile. *Z. Pflanzenernahr Dung. Bodenkd.*, **67**, 193–7.

Hamdi, H. and Naga, M. (1950) Chemical and mineralogical investigations of Egyptian soils. *Schweizer Mineralog. Petrog. Mitt.*, **29**, 537–40.

Hanna, F.S. and Beckmann, H. (1975) Clay minerals of some soils of the Nile valley in Egypt. *Geoderma*, **14**, 159–70.

Hanna, W.E., Daugherty, L.A. and Arnold, R.W. (1975) Soil-geomorphic relationships in a first-order valley in central New York. *Proceedings of the Soil Science Society of America*, **39**, 716–22.

Harden, J.W. (1982) A quantitative index of soil development from field descriptions: Examples from a chronosequence in central California. *Geoderma*, **28**, 1–28.

Harradine, F. and Jenny, H. (1958) Influence of parent material and climate on texture and nitrogen and carbon contents of virgin California soils. *Soil Science*, **85**, 235–43.

Harris, C. (1977) Engineering properties, groundwater conditions, and the nature of soil movement on a solifluction slope in north Norway. *Quarterly Journal of Engineering Geology*, **10**, 27–43.

Harris, C. (1981a) Microstructures in solifluction sediments from south Wales and north Norway. *Biuletyn Peryglacjalny*, **28**, 221–6.

Harris, C. (1981b) *Periglacial mass wasting: a review of research*, British Geomorphological Research Group, Research Monograph, 4, Geo Books, Norwich.

Harris, C. (1983) Vesicles in thin sections of periglacial soils from north and south Norway. *Proceedings of the 4th International Conference on Permafrost, Fairbanks, Alaska*, 445–9.

Harris, C. (1985) Geomorphological applications of soil micromorphology with particular reference to periglacial sediments and processes, in *Geomorphology and Soils* (eds K.S. Richards, R.R. Arnett and S. Ellis), Allen and Unwin, London, pp. 219–32.

Harris, C. and Ellis, S. (1980) Micromorphology of soils in soliflucted materials, Okstindan, northern Norway. *Geoderma*, **23**, 11–29.

Harris, S.A. (1958) The gilgaied and bad-structured soils of Central Iraq. *Journal of Soil Science*, **9**, 169–85.

Harris, S.A. (1980) Climatic relationship of permafrost in areas of low winter snow cover. *Biuletyn Peryglacjalny*, **28**, 227–40.

Harris, S.A. (1982) Distribution of zonal permafrost landforms with freezing and thawing indices. *Erdkunde*, **35**, 81–90.

Harrison, S.C. (1975) Tidal-flat complex, Delmarva Peninsula, Virginia, in *Tidal deposits: a casebook of recent examples and fossil counterparts* (ed R.N. Ginsburg), Springer-Verlag, New York.

Hashad, M.N. and Mady, F. (1961) Differential thermal analysis of the alluvial clay of Egypt. *Journal of Soil Science (UAR)*, **1**, 125–39.

Havinga, A.J. (1969) A physiographic analysis of a part of the Betuwe, a Dutch river clay area. *Mededelingen Landbouwhogeschool Wageningen, Nederland*, **3**, 69.

Havinga, A.J. and Opt'Hof, A. (1983) Physiography and formation of the Holocene floodplain along the lower course of the Rhine in the Netherlands. *Mededelingen Landbouwhogeschool Wageningen, Nederland*, **83–8**.

Hawker, H.W. (1927) A study of the soils of Higalgo County, Texas and the stages of their lime accumulation. *Soil Science*, **23**, 475–83.

Hawkins, D.M. and Merriam, D.F. (1973) Optimal zonation of digitized sequential data. *Mathematical Geology*, **5**, 389–95.

Hawkins, D.M. and Merriam, D.F. (1974) Zonation of multivariate sequences of geological data. *Mathematical Geology*, **6**, 263–9.

Hawley, J.W., Kottlowski, F.E., Strain, W.S. *et al.* (1969) The Sante Fe Group in the south-central New Mexico border region, in *Border Stratigraphy Symposium*, New Mexico Bureau of Mines and Mineral Resources, Circular 104, 52–76.

Hayward, M. (1985) Soil development in Flandrian floodplains: River Severn case study, in *Soils and Quaternary Landscape Evolution* (ed J. Boardman), Wiley, Chichester, pp. 281–99.

Hayward, M. and Fenwick, I. (1983) Soils and hydrological change, in *Background to Palaeohydrology* (ed K.J. Gregory), Wiley, Chichester, pp. 167–87.

Heede, B.H. (1974) Stages of development of gullies in the western United States. *Zeitschrift für Geomorphologie*, **19**, 260–71.

Heine, K. (1972) Die Bedeuting pedoligischer Untersuchungen bei der Trennung von Reliefgenerationen. *Zeitschrift für geomorphologie*, **14**, 113–37.

Herath, J.W. (1962) The mineralogical composition of some Ceylon Dry Zone clays. *Tropical Agriculturalist*, **118**, 47–54.

Herath, J.W. and Grimshaw, R.W. (1971) A general evaluation of the frequency distribution of clay and associated minerals in the alluvial soils of Ceylon. *Geoderma*, **5**, 119–30.

Herbel, C.H. and Gile, L.H. (1973) Field moisture regimes and morphology of some arid-land soils in New Mexico. *Soil Science Society of America Special Publication*, **5**, 119–52.

Hewitt, K. (1968) The freeze-thaw environment of the Karakoram Himalaya. *Canadian Geographer*, **12**, 85–98.

Hewlett, J.D. and Nutter, W.L. (1970) *The varying source area of streamflow from upland basins*. Paper presented at Symposium on Interdisciplinary Aspects of

Watershed Management, Montana State University, Bozeman, New York, American Society of Civil Engineers.

Hill, D.E. and Tedrow, J.C.F. (1961) Weathering and soil formation in the arctic environment. *American Journal of Science*, **259**, 84–101.

Hoeksema, K.J. (1953) The natural homogenisation of the soil profile in the Netherlands. *Boor en Spade*, **6**, 24–30.

Hole, F.D. (1976) *The sols of Wisconsin*, University of Wisconsin Press, Wisconsin.

Holmes, C.D. (1955) Geomorphic development in humid and arid regions: a synthesis. *American Journal of Science*, **253**, 377–90.

Holmes, D.A. and Western, S. (1969) Soil-texture patterns in the alluvium of the lower Indus Plains. *Journal of Soil Science*, **20**, 23–37.

Holmes, G.W., Hopkins, D.M. and Foster, H.J. (1968) Pingos in central Alaska. *United States Geological Survey, Bulletin*, **1241–H**, 40pp.

Hoover, M.T. and Ciolkosz, E.J. (1988) Colluvial soil parent material relationships in the Ridge and Valley Physiographic Province of Pennsylvania. *Soil Science*, **145**, 163–72.

Horton, R.E. (1945) Erosional development of streams and their drainage basins: hydrophysical approach to quantitative morphology. *Bulletin of the Geological Society of America*, **56**, 275–370.

Huddleston, J.H. and Riecken, F.F. (1973) Local soil-landscape relationships in Western Iowa. I. Distribution of selected chemical and physical properties. *Proceedings of the Soil Science Society of America*, **37**, 264–70.

Huggett, R.J. (1973) *The theoretical behaviour of materials within soil landscape systems*. University College, London, Department of Geography, Occasional Paper, 19.

Huggett, R.J. (1975) Soil landscape systems: a model of soil genesis. *Geoderma*, **13**, 1–22.

Hughes, O. (1969) *Distribution of open system pingos in central Yukon Territory with respect to glacial limits*. Geological Survey of Canada, Paper, 69–34, 8pp.

Hughes, O.L., Van Evendingen, R.D. and Tarnocai, C. (1983) Regional setting physiography and geology, in *Guidebook to permafrost and related features of the northern Yukon Territory and Mackenzie Delta, Canada* (eds H.M. French and J.A. Heginbottom), Guidebook 3, 4th International Conference of Permafrost, Fairbanks, Alaska, pp. 5–34.

Hutcheson, T.B., Lewis, R.J. and Seay, W.A. (1959) Chemical and clay mineralogical properties of certain Memphis catena soils of western Kentucky. *Proceedings of the Soil Science Society of America*, **23**, 474–8.

Hutchinson, J.N. (1974) Discussion to engineering problems caused by fossil permafrost features in the English Midlands, by A.V. Morgan. *Quarterly Journal of Engineering Geology*, **7**, 100.

Hutton, J.T. (1968) The redistribution of the more soluble chemical elements associated with soils as indicated by analysis of rainwater, soils and plants. *Transactions of the 9th International Congress of Soil Science, Adelaide, Australia*, **4**, 313–12.

Hutton, J.T. and Leslie, T.I. (1958) Accession of non-nitrogenous ions dissolved in rainwater to soils in Victoria. *Australian Journal of Agricultural Research*, **9**, 492–507.

Ignatenko, I.V. (1963) Arctic tundra soils of the Yugor Peninsula. *Soviet Soil Science*, **5**, 429–40.

Ignatenko, I.V. (1967) Soil complexes in Vaygach Island. *Soviet Soil Science*, **9**, 1216–29.

Islam, M.A. (1966) Soils of East Pakistan, in *Scientific Problems of the Humid Tropical Zone: Deltas and their implications*, UNESCO, New York, pp. 83–7.

Ismail, F.T. (1970) Biotite weathering and clay formation in arid and humid regions, California. *Soil Science*, **109**, 257–61.

Ives, D.W. and Cutler, E.J.B. (1972) A toposequence of steepland soils in the drier high-country yellow-brown earth (dry-hygrous elderfolvic) region, Canterbury, New Zealand. *New Zealand Journal of Science*, **15**, 385–407.

Jackson, L.E., Jr., McDonald, G.M. and Wilson, M.C. (1982) Paraglacial origin for terraced river sediments in Bow Valley, Alberta. *Canadian Journal of Earth Science*, **19**, 2219–31.

Jackson, M.L., Tyler, S.A., Bourbeau, G.A. *et al.* (1948) Weathering sequence of clay-size minerals in soils and sediments. I. Fundamental generalisations. *Journal of Physical and Colloid Chemistry*, **52**, 1237–60.

Jacobsen, T. and Adams, R.M. (1958) Salt and silt in ancient Mesopotamian agriculture. *Science*, **128**, 1251–8.

Jacobson, G.L. and Birks, H.J.B. (1980) Soil development on recent end moraines of the Klutlan glacier, Yukon Territory, Canada. *Quaternary Research*, **14**, 87–100.

Jahn, A. (1968) Denudational balance of slopes. *Geographica Polonica*, **13**, 9–29.

James, P.A. (1970) The soils of the Rankin Inlet area, Keewatin, NWT, Canada. *Arctic and Alpine Research*, **2**, 293–302.

Jelgersma, S. (1961) Holocene sea level changes in the Netherlands. *Meded. Geol. Sticht, Ser C VI*, **7**, 1–100.

Jenkins, D.A. (1985) Chemical and mineralogical composition in the identification of palaeosols, in *Soils and Quaternary Landscape Evolution* (ed J. Boardman), Wiley, Chichester, pp. 23–43.

Jenny, H. (1941) *Factors of soil formation. A system of quantitative pedology*, McGraw-Hill, New York.

Jenny, H. (1961a) Derivation of state factor equations of soils and ecosystems. *Proceedings of the Soil Science Society of America*, **25**, 385–8.

Jenny, H. (1961b) *E.W. Hilgard and the birth of modern soil science*, Forallon Publishers, Pisa, Italy and Berkeley, California.

Jessup, R.W. (1960a) The Stony Tableland soils of the southeastern portion of the Australian Arid Zone and their evolutionary history. *Journal of Soil Science*, **11**, 188–96.

Jessup, R.W. (1960b) Identification and significance of buried soils of Quaternary age in the southeast portion of the Australian arid zone. *Journal of Soil Science*, **12**, 199–213.

Johnson, D.L. and Rockwell, T.K. (1982) Soil geomorphology: theory, concepts and principles with examples and applications on alluvial and marine terraces in coastal California. *Geological Society of America, Programs with Abstracts*, **14**, 176.

Johnson, D.L. and Watson-Stegner, D. (1987) Evolution model of pedogenesis. *Soil Science*, **143**, 349–66.

Johnson, D.L., Keller, E.A. and Rockwell, T.K. (1990) Dynamic pedogenesis: New views on some key soil concepts and a model for interpreting Quaternary soils, *Quaternary Research*, **33**, 306–9.

Jones, R.L. and Beavers, A.H. (1964) Variation of opal phytolith content among some great soil groups of Illinois. *Proceedings of the Soil Science Society of America*, **28**, 711–12.

Jordan, M.K., Behling, R.E. and Kite, J.S. (1987) Characterization and late Quaternary history of the alluvial and colluvial deposits of the Pendleton Creek Valley, Tucker County, West Virginia. *Geological Society of America, Programs with Abstracts*, **19**, 92.

Jordan, S. (1974) The variability of soil in a Cotswold Valley, unpublished B.Sc. dissertation, School of Geography, The University of Birmingham.

Joseph, K.T. (1968) A toposequence of limestone parent material in north Kedah, Malaya. *Journal of Tropical Geography*, **27**, 19–22.

Kachurin, S.P. (1959) *Principles of geocryology. Part I. General cryology*, Academy of Sciences of USSR, V.A. Obruchev Institute of Permafrost Studies, National Research Council of Canada, Translation TT-1157, 1964.

Kamps, L.F. (1962) Mud distribution and land reclamation in the eastern Wadden shallows. *Rijkswat St. Commun.*, **4**, 1–73.

Karageorgis, D., Tonkin, P.J. and Adams, J.A. (1984) Medium and short range variability in textural layering in an Ochrept developed on an alluvial floodplain. *Australian Journal of Soil Research*, **22**, 471–4.

Karavayeva, N.A. and Targul'yan, V.O. (1960) Humus distribution in the tundra soils of northern Yakutia. *Soviet Soil Science*, **12**, 1293–300.

Kay, G.F. (1916) Gumbotil, a new term in Pleistocene geology. *Science*, **44**, 637–8.

Kerney, M.P. (1963) Late-glacial deposits on the Chalk of southeast England. *Philosophical Transactions of the Royal Society*, **B246**, 203–54.

Kesel, R.H., Dunne, K.C., McDonald, R.C. *et al.* (1974) Lateral erosion and overbank deposition in the Mississippi River in Louisiana caused by 1973 flooding. *Geology*, **2**, 461–4.

Khadr, M. (1960) The clay mineral composition of some soils of the UAR. *Journal of Soil Science, UAR*, **1**, 141–55.

King, L.C. (1953) Canons of landscape evolution. *Bulletin of the Geological Society of America*, **64**, 721–51.

Kirkby, M.J. (1969) Erosion by water on hillslopes, in *Water, Earth and Man* (ed R.J. Chorley), Methuen, London, pp. 229–38.

Kirkham, D. (1965) Seepage of leaching water into drainage ditches of unequal water level heights. *Journal of Hydrology*, **3**, 207–24.

Kline, J.R. (1973) Mathematical simulation of soil-plant relationships and soil genesis. *Soil Science*, **115**, 240–49.

Knight, M.J. (1975) Recent crevassing of the Erap River, Papua New Guinea. *Australian Geographical Studies*, **13**, 77–84.

Knuepfer, P.L.K. and McFadden, L.D. (eds) (1990) *Soils and landscape evolution*, Elsevier, Amsterdam.

Kobayashi, K. (1965) Late Quaternary chronology of Japan. *Earth Science, Japan*, **79**, 1–17.

Kochel, R.C. and Baker, V.R. (1982) Paleoflood hydrology. *Science*, **215**, 353–61.

Koniscev, V.N., Faustova, M.A. and Rogov, V.V. (1973) Cryogenic processes as reflected in ground microstructures. *Biuletyn Peryglacjalny*, **22**, 213–19.

Kovda, V.A., Lobova, Ye V. and Rozanov, V.V. (1967) Classification of the world's soils. *Soviet Soil Science*, **2**, 851–63.

Kreida, N.A. (1958) Soils of the eastern European tundras. *Soviet Soil Science*, **1**, 51–6.

Krusekopf, H.H. (1948) Gumbotil – its formation and relation to overlying soils with clay pan subsoils. *Proceedings of the Soil Science Society of America*, **12**, 413–14.

Kubiena, W.L. (1938) *Micropedology*, Collegiate Press, Ames, Iowa.

Kubiena, W.L. (1953) *The soils of Europe*, Thomas Murby, London.

Kwad, F.J.P.M. and Mucher, H.J. (1977) The evolution of soils and slope deposits in the Luxembourg Ardennes, near Wiltz. *Geoderma*, **17**, 1–37.

Kwad, F.J.P.M. and Mucher, H.J. (1979) The formation and evolution of colluvium on arable land in northern Luxembourg. *Geoderma*, **22**, 173–92.

Lafeber, D. (1965) The graphical representation of planar pore patterns in soils. *Australian Journal of Soil Research*, **3**, 143–64.

Lafeber, D. (1966) Soil structural concepts. *Engineering Geology*, **4**, 261–90.

Lanyon, L.E. and Hall, G.F. (1983) Land-surface morphology. 2. Predicting landscape instability in Eastern Ohio. *Soil Science*, **136**, 382–6.

Larsen, G. and Thorarinsson, S. (1978) H4 and other acid Hekla tephra layers. *Jökull*, **27**, 28–46.

Leamy, M.L. and Burke, A.S. (1973) Identification and significance of paleosols in cover deposits in Central Orago. *New Zealand Journal of Geology and Geophysics*, **16**, 623–35.

Leighton, M.M. and MacClintock, P. (1962) The weathered mantle of glacial tills beneath original surfaces in north-central United States. *Journal of Geology*, **70**, 267–93.

Leo, R.F. and Barghoorn, E.S. (1976) Silica in the biosphere. *Acta Cientifica Venezolana*, **27**, 231–4.

Leopold, L.B. and Wolman, M.G. (1957) *River channel patterns – braided, meandering and straight*, United States Gelogical Survey, Professional Paper, 282D, 39–85.

Leopold, L.B., Wolman, M.G. and Miller, J.P. (1964) *Fluvial processes in geomorphology*, W.H. Freeman, San Francisco.

Leslie, D.M. (1973) Quaternary deposits and surfaces in a volcanic landscape on Otago peninsula. *New Zealand Journal of Geology and Geophysics*, **16**, 557–66.

Lewin, J. (1978) Floodplain geomorphology. *Progress in Physical Geography*, **2**, 408–37.

Lewis, W.V. (1932) The formation of Dungeness foreland. *Geographical Journal*, **80**, 309–24.

Lewis, W.V. and Balchin, W.G.V. (1940) Past sea-levels at Dungeness. *Geographical Journal*, **96**, 258–85.

Libby, W.F. (1955) *Radiocarbon dating*, 2nd edn, Chicago University Press, Chicago.

Liestøl, O. (1977) Pingos, springs and permafrost in Spitsbergen. *Norsk Polarinstitutt, Arbok, 1975*, Oslo, 7–29.

Limmer, A.W. and Wilson, A.T. (1980) Amino acids in buried paleosols. *Journal of Soil Science*, **31**, 147–53.

Lipscomb, G.H. and Farley, W.H. (1981) *Soil survey of Juniata and Mifflin Counties, Pennsylvania*, United States Department of Agriculture, Soil Conservation Service, Washington, D.C., 177pp.

Lobova, E.V. (1967) *Soils of the desert zone of the U.S.S.R.*, (translated) Israel Programme for Scientific Translations, Jerusalem.

Lorenzo, J. (1969) Minor periglacial phenomena among the high volcanoes of Mexico, in *The periglacial environment* (ed T.L. Pewe), McGill-Queens University Press, Montreal, pp. 161–75.

Losche, C.K., McCracken, R.J. and Davey, C.B. (1970) Soils of steeply sloping landscapes in the Southern Appalachian Mountains. *Proceedings of the Soil Science Society of America*, **34**, 473–8.

Lotspeich, F.B. and Smith, H.W. (1953) Soils of the Palouse loess: I. The Palouse catena. *Soil Science*, **76**, 467–80.

Louis, H. (1964) Uber Rumpfflachen und Talbilding in den wechselfenchten Tropen besonders nach Studien in Tanganyka. *Zeitschrift für geomorphologie*, **8**, 43–70.

Lozinski, W. von (1909) Uber die mechanische Verwitterung der Sandsteine im gemassigten klima. *Acad. Sci. Cracovie Bull. Internat. Cl. Sci., Math et Nat.*, **1**, 1–25.

Lozinski, W. von (1912) Die periglaziale Fazies der mechanischen Verwitterung, *Comptes Rendus, XI Congres Internationale Geologie, Stockholm, 1910*, 1039–53.

Lundqvist, J. (1962) Patterned ground and related frost phenomenon in Sweden, *Sveriges Geologie Unders Ser. C583*, Stockholm.

Mabbutt. J.A. (1969) Landforms of arid Australia, in *Arid lands of Australia* (eds R.O. Slatyer and R.A. Perry), Australian National University Press, Canberra, pp. 11–32.

Mabbutt, J.A. (1971) The Australian arid zone as a prehistoric environment, in *Aboriginal man and environment in Australia* (eds D.J. Mulvaney and J. Golson), Australian National University Press, Canberra, pp. 66–79.

Mabbutt, J.A. and Scott, R.M. (1966) Periodicity of morphogenesis and soil formation in a Savannah landscape near Port Moresby, Papua. *Zeitschrift für Geomorphologie*, **10**, 69–89.

Machette, M.N. (1985) Calcic soils of the southwestern united States. *Geological Society of America, Special Paper*, **203**, 1–21.

McArthur, W.M. and Bettenay, E. (1960) *The development and distribution of the soils of the Swan coastal plain Western Australia*, CSIRO, Soil Publication 16.

McBratney, A.B. and Webster, R. (1981) Spatial dependence and classification of the soil along a transect in North East Scotland. *Geoderma*, **26**, 63–82.

MacClaren, M. (1906) On the origin of certain laterites. *Geological Magazine*, **43**, 536–47.

McCraw, J.D. (1968) The soil pattern of some New Zealand alluvial fans. *Transactions of the 9th International Congress of Soil Science*, **4**, 631–40.

McFadden, L.D. and Knuepfer, P.L.K. (1990) Soil geomorphology – the linkage of pedology and surficial processes. *Geomorphology*, **3**, 197–205.

McFadden, L.D. and Tinsley, J.C. (1985) The rate and depth of accumulation of pedogenic carbonate accumulation in soils: Formation and testing of a compartment model, in *Soils and Quaternary geology of the southwestern United States* (ed D.W. Weide), Geological Society of America, Special Paper, 203, 23–42.

MacFarlane, M.J. (1971) Laterization and landscape development in Kyagive, Uganda. *Quarterly Journal of the Geological Society, London*, **126**, 501–39.

MacFarlane, M.J. (1973) Laterite and topography in Buganda. *Uganda Journal*, **36**, 9–22.

MacFarlane, M.J. (1976) *Laterite and Landscape*, Academic Press, London.

MacFarlane, M.J. (1983) Laterites, in *Chemical sediments and geomorphology* (eds A.S. Goudie and K.Pye), Academic Press, London, pp. 7–58.

McGee, W.J. (1891) *The Pleistocene history of northeastern Iowa*, United States Geological Survey Report, 11, 189–577.

Mackay, J.R. (1962) Pingos of the Pleistocene Mackenzie River Delta area. *Geographical Bulletin*, **18**, 21–63.

Mackay, J.R. (1973) The growth of pingos, Western Arctic coast, Canada. *Canadian Journal of Earth Sciences*, **10**, 979–1004.

Mackay, J.R. (1979) An equilibrium model for hummocks (non-sorted circles), Garry Island, Northwest territories. *Canadian Geological Survey, Paper*, **79-1A**, 165–7.

Mackay, J.R. (1980) The origin of hummocks, western Arctic coast, Canada. *Canadian Journal of Earth Sciences*, **17**, 996–1006.

Mackay, J.R., Mathews, W.H. and Nacneish, R.S. (1961) Geology of the Engigstciak archaeological site, Yukon Territory. *Arctic*, **14**, 25–52.

McKeague, J.A., Acton, C.J. and Dumanski, J. (1974) Studies of soil micromorphology in Canada, in *Soil microscopy* (ed G.K. Rutherford), Limestone Press, Kingston, pp. 84–100.

McKeague, J.A., Ross, G.J. and Gamble, D.S. (1978) Properties, criteria of classification and concepts of genesis of podzolic soils in Canada, in *Quaternary Soils* (ed W.C. Mahaney), Geo Books, Norwich.

McKellar, I.C. (1960) Pleistocene deposits of the Upper Clutha Valley, Otago, New Zealand. *New Zealand Journal of Geology and Geophysics*, **3**, 432–60.

McLellan, A.G. (1971) Some economic implications of research methods used in glacial geomorphology, *Research methods in geomorphology, Proceedings of the 1st Guelph Symposium*, 57–72.

McMillan, N.J. (1960) Soils of the Queen Elizabeth Islands (Canadian Arctic). *Journal of Soil Science*, **11**, 131–9.

Mahaney, W.C. (1978a) *Quaternary Soils*, Geo Books, Norwich.

Mahaney, W.C. (1978b) Late Quaternary stratigraphy and soils in the Wind River Mountains, Western Wyoming, in *Quaternary Soils*, Geo Books, Norwich. pp. 223–64.

Marinkovic, P. (1964) Classification of alluvial soils. *Transactions of the 8th International Congress of Soil Science*, **5**, 675–8.

Markewich, H.W. (1985) Geomorphic evidence for Pliocene-Pleistocene uplift in the area of the Cape Fear Arch, North Carolina, in *Tectonic Geomorphology* (eds M Morisawa and J.T. Hack). Proceedings of the 15th Annual Binghamton Geomorphology Symposium, Allen and Unwin, London, pp. 279–98.

Markewich, H.W., Pavich, M.J. and Buell, G.R. (1990) Contrasting soils and landscapes of the Piedmont and Coastal Plain, eastern United States. *Geomorphology*, **3**, 417–47.

Markewich, H.W., Pavich, M.J., Mausbach, M.J. *et al.* (1986) Soil development and its relation to the ages of morphostratigraphic units in Horry County, South Carolina. *United States Geological Survey Bulletin*, **1589-B**, 61pp.

Markewich, H.W., Pavich, M.J., Mausbach, M.J. *et al.* (1987) Age relations between soils and geology in the Coastal Plain of Maryland and Virginia. *United States Geological Survey Bulletin*, **1589-A**, 34pp.

Markewich, H.W., Pavich, M.J. Mausbach, M.J. *et al.* (1989) A guide for using soil and weathering profile data in chronosequence studies of the Coastal Plain of the eastern United States. *United States Geological Survey Bulletin*, **1589-D**, 39pp.

Markos, G. (1977) Geochemical alteration of plagioclase and biotite in glacial and periglacial deposits, unpublished PhD. thesis, University of Colorado, Boulder.

Martel, Y.A. and Paul, E.A. (1974) The use of radiocarbon dating for organic matter in the study of soil genesis. *Proceedings of the Soil Science Society of America*, **38**, 501–6.

Matthes, F.E. (1900) Glacial sculpture of the Bighorn Mountains, Wyoming. *United States Geological Survey, 21st Annual Report, 1899–1900*, 167–90.

Matthews, B. (1971) Soils in Yorkshire. I. Sheet SE65 (York East). *Soil Survey Record*, **6**, Soil Survey, Harpenden.

Matthews, J.A. (1980) Some problems and implications of ^{14}C dates from a podzol buried beneath an end moraine at Haugabreen, southern Norway. *Geografiska Annaler*, **62A**, 185–208.

Matthews, J.A. (1981) Natural ^{14}C age/depth gradient in a buried soil. *Naturwissenschaften*, **68**, 472–4.

Matthews, J.A. (1984) Limitations of ^{14}C dates from buried soils in reconstructing glacier variations and Holocene climate, in *Climatic changes on a yearly to millenial basis: geological, historical and instrumental records* (eds N.A. Morner and W. Karlen), Reidel, Dordrecht, pp. 281–90.

Matthews, J.A. (1985) Radiocarbon dating of surface and buried soils: principles, problems and prospects, in *Geomorphology and Soils* (eds K.S. Richards, R.R. Arnett and S.Ellis), Allen and Unwin, London, pp. 269–88.

Matthews, J.A. and Dresser, P.Q. (1983) Intensive ^{14}C dating of a buried palaeosol horizon. *Geol. Foreningens*, **195**, 59–63.

Mayland, H.F. and McIntosh, T.H. (1963) Nitrogen fixation by desert algal crust organisms. *Agronomy Abstracts*, **1963**, 33.

Meigs, P. (1953) World distribution of arid and semi-arid homoclimates, in *Reviews of Research in Arid Zone Hydrology*, UNESCO, Paris, pp. 203–9.

Meixner, R.E. and Singer, M.J. (1981) Use of a field morphology rating system to evaluate soil formation and discontinuities. *Soil Science*, **131**, 114–23.

Mellor, A. (1985) Soil chronosequences on neoglacial moraine ridges, Jostedalsbreen and Jotunheimen, southern Norway: A Quantitative pedogenic approach, in *Geomorphology and Soils* (eds K.S. Richards, R.R. Arnett and S.Ellis), Allen and Unwin, London, pp. 289–308.

Merkel, E.J. (1978) Soil survey of Huntington County, Pennsylvania, *United States Department of Agriculture, Washington, D.C.*, 122pp.

Miall, A.D. (1977) A review of the braided river depositional environment. *Earth Science Review*, **13**, 1–62.

Middleton, H.E. (1930) *Properties of soils which influence soil erosion*, United States Department of Agriculture, Technical report, 178, 1–16.

Middleton, N. (1991) *Desertification*, Oxford University Press, Oxford.

Milne, G. (1935a) Some suggested units of classification and mapping particularly for East African soils. *Soils Research*, **4**, no. 3.

Milne, G. (1935b) Composite units for the mapping of complex soil associations. *Transactions of the 3rd International Congress of Soil Science*, **1**, 345–7.

Milne, G. (1936a) Normal erosion as a factor in soil profile development. *Nature*, **138**, 548–9.

Milne, G. (1936b) *A provisional soil map of East Africa*, East Africa Agricultural Research Station, Amani, Memoir, 34.

Milne, G. (1947) A soil reconnaissance journey through parts of Tanganyika Territory, December 1935 to February 1936. *Journal of Ecology*, **35**, 192–265.

Moar, N.T. (1970) A new pollen diagram from Pyramid Valley swamp. *Records Canterbury Museum*, **8**, 455–61.

Mohr, E.C. and Van Baren, F.A. (1954) *Tropical soils*, N.V. Vitgeverij W. van Hoeve, the Hague.

Moore, T.R. (1976) Sesquioxide-cemented soil horizons in northern Quebec: their distribution, properties and genesis. *Canadian Journal of Soil Science*, **56**, 333–44.

Morin, J. and Jarosch, H.S. (1977) Rainfall-runoff analysis for bare soils, *Pamphlet*, **164, 3**, Volcanic Center, Division of Scientific Publications, Bet Daga, Israel, 22pp.

Morozova, T.D. (1965) Micromorphological characteristics of pale yellow permafrost soils in central Yakutia in relation to cryogenesis. *Soviet Soil Science*, **7**, 1333–42.

Morrison, R.B. (1964) *Lake Lahontan: Geology of southern Carson Desert, Nevada*. US Geological Survey Professional Paper, No. 401.

Morrison, R.B. (1965) *Lake Bonneville. Quaternary stratigraphy of eastern Jordan Valley, south of Salt Lake City, Utah*. US Geological Survey Professional Paper, No. 477.

Morrison, R.B. (1967) Principles of Quaternary soil stratigraphy, in *Quaternary Soils* (eds R.B. Morrison and H.E. Wright, Jr.), INQUA 7th Congress 1965, Proceedings no. 9, University of Nevada, pp. 1–69.

Morrison, R.B. (1978) Quaternary soil stratigraphy – concepts, methods and problems, in *Quaternary Soils* (ed W.C. Mahaney), Geo Books, Norwich, pp. 77–108.

Morrison, R.B. and Wright, H.E. (eds) (1967) *Quaternary Soils*, INQUA 7th Congress, 1965, Proceedings no. 9, University of Nevada.

Mosley, M.P. (1978) Erosion in the south-eastern Ruahine Range: its implications for downstream river control. *New Zealand Journal of Forestry*, **23**, 21–48.

Moss, A.J. (1962) The physical nature of common sandy and pebbly deposits. *American Journal of Science*, **260**, 337–73.

Moss, A.J. (1963) The physical nature of common sandy and pebbly deposits, II. *American Journal of Science*, **261**, 297–343.

Moss, A.J. (1972) Bed-load sediments. *Sedimentology*, **18**, 159–219.

Moss, J.H. (1977) The formation of pediments – scarp backwearing or scarp downwasting? in *Geomorphology of arid regions* (ed D.O. Doehring), State University of New York, Binghamtom, pp. 51–78.

Moss, J.H. and Kochel, R.C. (1978) Unexpected geomorphic effects of the Hurricane Agnes storm and flood, Conestoga Drainage Basin, Pennsylvania. *Journal of Geology*, **86**, 1–11.

Moss, R.P. (1963) Soil, slopes and landuse in a part of south-west Nigeria: some implications for the planning of agricultural development in inter-tropical Africa. *Transactions of the Institute of British Geographers*, **32**, 143–68.

Moss, R.P. (1965) Slope development and soil morphology in a part of south-west Nigeria. *Journal of Soil Science*, **16**, 192–209.

Moss, R.P. (1968) Soils, slopes and surfaces in tropical Africa, in *The soil resources of tropical Africa*, Cambridge University Press, Cambridge, pp. 29–60.

Motts, W.S. (1958) Caliche genesis and rainfall in the Pecos Valley area of southeastern New Mexico. *Bulletin of the Geological Society of America*, **69**, 1737.

Muckenhirn, R.J., Whiteside, E.P., Templin, E.H. *et al.* (1949) Soil classification and the genetic factors of soil formation. *Soil Science*, **67**, 93–105.

Muhs, D.R. (1982) A soil chronosequence on Quaternary marine terraces, San Clemente Island, California. *Geoderma*, **28**, 257–83.

Mulcahy, M.J. (1960) Laterites and lateritic soils in south-western Australia. *Journal of Soil Science*, **11**, 206–26.

Mulcahy, M.J. (1961) Soil distribution in relation to landscape development. *Zeitschrift für Geomorphologie, Supplementband*, **5**, 211–25.

Muller, S.W. (1945) Permafrost or perennially frozen ground and related engineering problems. *United States Geological Survey Special Report Strategic Engineering Study*, **62**, 231pp.

Nabhan, H.M., Sys, C. and Stoops, G. (1969) Mineralogical study on the suspended matter in the Nile water. *Pedologie*, **29**, 34–48.

Nanson, G.C. (1980) Point bar and floodplain formation of the meandering Beatton River, northeast British Columbia, Canada. *Sedimentology*, **27**, 3–29.

Nash, D. (1980) Forms of bluffs degraded for different lengths of time in Emmet County, Michigan USA. *Earth Surface Processes*, **5**, 331–45.

Netterberg, F. (1980) Geology of southern African calcretes. I. Terminology, description, microfeatures and classification. *Transactions of the Geological Society of South Africa*, **83**, 255–83.

Nettleton, W.D., Flach, K.W. and Brasher, B.R. (1969) Argillic horizons without clay skins. *Proceedings of the Soil Science Society of America*, **33**, 121–5.

Nettleton, W.D., Witty, J.E., Nelson, R.E. *et al.* (1975) Genesis of argillic horizons in soils of desert areas of the southwestern United States. *Proceedings of the Soil Science Society of America*, **39**, 919–26.

Newill, D. (1961) A laboratory investigation of two red clays from Kenya. *Geotechnique*, **11**, 302–18.

Nikiforoff, C.C. (1937) General trends of the desert type of soil formation. *Soil Science*, **43**, 105–31.

Nikiforoff, C.C. (1949) Weathering and soil formation. *Soil Science*, **67**, 219–30.

Norton, E.A. and Smith, R.S. (1930) Influence of topography on soil profile character. *Journal of the American Society of Agronomists*, **22**, 251–62.

Nye, P.H. (1954–5) Some soil forming processes in the humid tropics. I-IV. *Journal of Soil Science*, **5**, 7–21 and 51–83.

Oberlander, T.M. (1974) Landscape inheritance and the pediment problem in the Mojave Desert of southern California. *American Journal of Science*, **274**, 849–75.

O'Brien, R.M.G., Romans, J.C.C. and Robertson, L. (1979) Three soil profiles from Elephant Island, South Shetland Islands. *Bulletin of the British Antarctic Survey*, **48**, 1–12.

O'Connor, K.F. (1980) The use of mountains: a review of New Zealand experience, in *The land our future: Essays on land use and conservation in New Zealand* (ed A.G. Anderson), Longman Paul, New Zealand Geographical Society, Wellington pp. 193–222.

O'Connor, K.F. (1984) Stability and instability of ecological systems in New Zealand Mountains. *Mountain Research and Development*, **4**, 15–29.

Oertel, A.C. (1968) Some observations incompatible with clay illuviation. *Transactions of the 9th International Congress of Soil Science*, **4**, 481–88.

Oertel, A.C. and Giles, J.B. (1966) Quantitative study of a layered soil. *Australian Journal of Soil Research*, **4**, 19–28.

Oliver, M.A. and Webster, R. (1987) The elucidation of soil pattern in the Wyre Forest of the West Midlands, England. II. Spatial distribution. *Journal of Soil Science*, **38**, 293–307.

Oliver, M.A., Webster, R. and Gerrard, A.J. (1989a) Geostatistics in physical geography. Part I. Theory. *Transactions of the Institute of British Geographers*, **NS 14**, 259–69.

Oliver, M.A., Webster, R. and Gerrard, A.J. (1989b) Geostatistics in physical geography. Part II. Applications. *Transactions of the Institute of British Geographers*, **NS 14**, 270–86.

Ollier, C.D. (1959) A two-cycle theory of tropical pedology. *Journal of Soil Science*, **10**, 137–48.

Ollier, C.D. (1976) Catenas in different climates, in *Geomorphology and climate* (ed E. Derbyshire), Wiley, Chichester.

Ollier, C.D. (1978) Silcrete and weathering, in *Silcrete in Australia* (ed T. Langford-Smith), University of New England, Armidale, Australia, pp. 13–17.

Ollier, C.D. and Thomasson, A.J. (1957) Asymmetrical valleys of the Chiltern Hills. *Geographical Journal*, **123**, 71–80.

O'Loughlin, C.L. (1969) Stream bed investigations in a small mountain catchment. *New Zealand Journal of Geology and Geophysics*, **12**, 684–706.

Olson, C.G. and Doolittle, J.A. (1985) Geophysical techniques for reconnaisance investigations of soils and surficial deposits in mountainous terrain. *Journal of the Soil Science Society of America*, **49**, 1490–98.

Olson, C.G. and Hupp, R. (1986) Coincidence and spatial variability of geology, soils and vegetation, Mill Run Watershed, Virginia. *Earth Surface Processes and Landforms*, **11**, 619–29.

Olsson, I.V. (1974) Some problems in connection with the evaluation of ^{14}C dates. *Geol. Foren. i Stockh. Forh.*, **96**, 311–20.

Ongley, E.D. (1970) Determination of rectilinear profile segments by automatic data processing. *Zeitschrift für Geomorphologie, Supplementband*, **14**, 383–91.

Oschwald, W.R. et al. (1965) *Principal soils of Iowa*, Iowa State University Extension Service Special report, no. 42.

Pallister, J.W. (1951) *Occurrence of laterite in south Buganda*. Report of the Geological Survey of Uganda, JWP/7.

Panabokke, C.R. (1959) A study of some soils in the dry zone of Ceylon. *Soil Science*, **87**, 67–74.

Parizek, E.J. and Woodruff, J. (1957) Description and origin of stone layers in soils of the southeastern States. *Journal of Geology*, **65**, 24–34.

Parsons, R.B. (1978) Soil-geomorphology relations in mountains of Oregon, U.S.A. *Geoderma*, **21**, 25–39.

Parsons, R.B., Balster, C.A. and Ness, A.O. (1970) Soil development and geomorphic surfaces, Willamette Valley, Oregon. *Proceedings of the Soil Science Society of America*, **34**, 485–91.

Patton, P.C. and Schumm, S.A. (1975) Gully erosion northwestern Colorado – a threshold phenomenon. *Geology*, **3**, 88–90.

Patton, P.C., Alexander, C.S. and Kramer, F.L. (1970) *Physical Geography*, Wandsworth Publishing Company, Belmont, California.

Pawluk, S. (1978) The pedogenic profile in the stratigraphic section, in *Quaternary soils*, (ed W.C. Mahaney), Geo Books, Norwich.

Pawluk, S. and Brewer, R. (1975) Micromorphological and analytical characteristics of some soils from Devon and King Christian Islands, NWT. *Canadian Journal of Earth Science*, **55**, 349–61.

Pavich, M.J. (1985) Appalachian Piedmont morphogenesis: weathering, erosion and Cenozoic uplift, in *Tectonic Geomorphology*, (eds M. Morisawa and J.T Hack), Proceedings of the 15th Annual Geomorphology Symposium, Binghamton, New York, Allen and Unwin, London, pp. 299–319.

Pavich, M.J. (1986) Processes and rates of saprolite production and erosion on a foliated granitic rock of the Virgina Piedmont, in *Rates of chemical weathering of rocks and minerals*, Academic Press, New York, pp. 551–90.

Pavich, M.J. (1989) Regolith residence time and the concept of surface age of the Piedmont 'peneplain'. *Appalachian Geomorphology*, **2**, 181–96.

Pavich, M.J., Brown, L., Valette-Silver, J.N. *et al.* (1985) [14]Be analysis of a Quaternary weathering profile in the Virginia Piedmont. *Geology*, **13**, 39–41.

Pavich, M.J., Leo, G.W., Obermeier, S.F. *et al.* (1989) Investigations of the characteristics, origin and residence time of the upland residual mantle of the Piedmont of Fairfax County, Virginia. *United States Geological Survey, Professional Paper*, **1352**, 58pp.

Peltier, L.C. (1950) The geographic cycle in periglacial regions as it is related to climatic geomorphology. *Annals of the Association of American Geographers*, **40**, 214–36.

Penck, W. (1924) *Die Morphologische Analyse* (translated by H. Czech and K.C. Boswell 1953), MacMillan, London.

Perrin, R.M.S. and Mitchell, C.W. (1969, 1971) An appraisal of physiographic units for predicting site conditions in arid areas. *MEXE Report 1111*, 2 vols. (vol. 1, 1969, vol. 2 1971).

Pettijohn, F.J. (1941) Persistence of heavy minerals and geological age. *Journal of Geology*, **49**, 610–25.

Pewe, T.L. (1969a) The periglacial environment, in *The periglacial environment*, (ed T.L. Pewe), McHill-Queens University Press, Montreal, pp. 1–11.

Pewe, T.L. (ed) (1969b) *The periglacial environment*, McGill-Queens University Press, Montreal.

Pierce, K.L., Obradovich, J.D. and Friedam, I (1976) Obsidian hydration dating and correlation of Bull Lake and Pinedale glaciations near West Yellowstone, Montana. *Bulletin of the Geological Society of America*, **87**, 703–10.

Pilgrim, A.T. (1972) *The identification of geomorphological and pedological criteria for the recognition and delimitation of landsurface units 5 and 6 in a semi-arid environment, Western Australia.* Paper presented to 1972 Conference of Australian Geographers, Canberra.

Poag, C.W. and Sevon, W.D. (1989) A record of Appalachian denudation in post rift Mesozoic and Cenozoic sedimentary deposits of the U.S. Middle Atlantic continental margin. *Geomorphology*, **2**, 119–57.

Ponnamperuma, F.N. (1972) The chemistry of submerged soils. *Advances in Agronomy*, **24**, 29–96.

Pons, L.J. and Zonneveld, I.S. (1965) *Soil ripening and soil classification; initial soil formation of alluvial deposits with a classification of the resulting soils.* Wageningen International Institute for land Reclamation and Improvement, Wageningen.

Powell, J.W. (1876) *Geology of the Uinta Mountains*, Washington.

Pregitzer, K.S., Barnes, B.V. and Lemme, G.D. (1983) Relationship of topography to soils and vegetation in an upper Michigan ecosystem. *Journal of the Soil Science Society of America*, **47**, 117–23.

Prest, V.K. (1968) *Nomenclature of moraines and ice-flow features as applied to the glacial map of Canada.* Geological Survey of Canada, Paper 67.

Price, R.J. (1973) *Glacial and fluvioglacial landforms*, Oliver and Boyd, Edinburgh.

Priesnitz, K. (1988) Cryoplanation, in *Advances in periglacial geomorphology* (ed M.J. Clark), Wiley, Chichester, pp. 49–68.

Pullan, R.A. and De Leeuw, P.N. (1964) *The land capability survey prepared for the Niger Dams Resettlement Authority*, Samaru, Zaria, Northern Nigeria Institute of Agricultural Research, Ahmadu Bello University, Soil Survey Bulletin.

Pye, K. (1987) *Aeolian dust and dust deposits*, Academic Press, London.

Quigley, R.M. and Ogunbadejo, T.A. (1976) Till geology, mineralogy and geotechnical behaviour, in *Glacial till: an interdisciplinary study* (ed R.F. Legget), Royal Society of Canada, Special Publication, 12.

Raad, A.T. and Protz, R. (1971) A new method for the identification of sediment stratification in soils of the Blue Springs Basin, Ontario. *Geoderma*, **6**, 23–41.

Radwanski, S.A. and Ollier, C.D. (1959) A study of an East African catena. *Journal of Soil Science*, **10**, 149–68.

Raeside, J.D. (1948) Some post-glacial climate changes in Canterbury and their effects on soil formation. *Transactions of the Royal Society of New Zealand*, **77**, 153–71.

Raeside, J.D. (1964) Loess deposits of the South Island, New Zealand and the soils formed on them. *New Zealand Journal of Geology and Geophysics*, **7**, 811–38.

Raup, H.M. (1965) The structure and development of turf hummocks in the Mesters Vig district, northeast Greenland. *Meddeleser om Gronland*, **166**.

Razzaq, A. and Herbillon, A.J. (1979) Clay mineralogical trend in alluvium-derived soils of Pakistan. *Pedologie*, **29**, 5–23.

Reeves, B.O.K. and Dormaar, J.F. (1972) A partial Holocene pedological and archaeological record from the southern Alberta Rocky Mountains. *Arctic and Alpine Research*, **4**, 325–36.

Reeves, C.C. and Suggs, J.D. (1964) Caliche of central south Llano Estacado, Texas. *Journal of Sedimentary Petrology*, **34**, 699–720.

Reger, R.D. and Pewe, T.L. (1976) Cryoplanation terraces: indicators of a permafrost environment. *Quaternary Research*, **6**, 99–109.

Reijne, A. (1961) On the contribution of the Amazon River to accretion of the coast of the Guianas. *Geologische Mijnbouw*, **NS23**, 219–26.

Retallack, G.J. (1984) Completeness of the rock and fossil record: estimates using fossil soils. *Paleobiology*, **10**, 59–78.

Retallack, G.J. (1990) *Soils of the past*, Harper Collins, London.

Richards, K.S., Arnett, R.R. and Ellis, S. (1985) *Geomorphology and soils*, Allen and Unwin, London.

Richmond, G.M. (1962) Quaternary stratigraphy of the La Sal Mountains, Utah. *United States Geological Survey Professional Paper*, **324**.

Richmond, G.M. (1965) Glaciation of the Rocky Moutains, in *The Quaternary of the United States* (eds H.E. Wright and D.G. Frey), Princeton University Press, Princeton, New Jersey, pp. 217–30.

Richmond, G.M. (1970) Comparison of the Quaternary stratigraphy of the Alps and Rocky Mountains. *Quaternary Research*, **1**, 3–28.

Richmond, G.M. (1973) *Geologic map of the Fremont Lake, South Quadrangle.* United States Geological Survey, Geologic Quadrangle Map, GQ - 1138

Richter, H., Haase, G. and Barthel, H. (1963) Die Goletzterrassen. *Petermanns Geographishe Mitteilungen*, **107**, 183–92.

Riecken, F.F. and Poetsch, E. (1960) Genesis and classification considerations of some prairie-formed soil profiles from local alluvium in Adair County, Iowa. *Proceedings of the Iowa Academy of Sciences*, **67**, 268–76.

Rieger, S. (1966) Dark well-drained soils of tundra regions in western Alaska. *Journal of Soil Science*, **17**, 264–73.

Rieger, S. (1974) Arctic soils, in *Arctic and Alpine Environments* (eds J.D. Ives and R.G. Barry), Methuen, London, pp. 749–69.

Ritchie, J.C., Hawks, P.H. and McHenry, H.R. (1975) Deposition rates in valleys determined using fallout Cesium 137. *Bulletin of the Geological Society of America*, **86**, 1128–30.

Ritter, D.F. (1988) Landscape analysis and the search for unity. *Bulletin of the Geological Society of America*, **100**, 160–71.

Ritter, D.F., Kinsey, W.F. and Kauffman, M.E. (1973) Overbank sedimentation in the Delaware River valley during the last 6000 years. *Science*, **179**, 374–5.

Robinson, S.W. (1949) *Soils, their origin, constitution and classification*, 3rd edn, Allen and Unwin, London.

Romanov, Y.E.A. (1974) Effect of trapped gas on some processes in soils. *Soviet Soil Science*, **4**, 222–8.

Romans, J.C.C., Robertson, L. and Dent, D.L. (1980) The micromorphology of young soils from south-east Iceland. *Geografiska Annaler*, **62A**, 93–103.

Ross, C.W., Mew, G. and Searle, P.L. (1977) Soil sequences on two terrace systems in the north Westland area, New Zealand. *New Zealand Journal of Science*, **20**, 231–44.

Rougerie, G. (1960) Le faconnement actuel des modeles en Cote d'Ivorie forestiere. *Memoire Institute Francais d'Afrique Noir*, no. 58.

Rozov, N.N. and Ivanova, E.N. (1967) Classification of soils of the USSR. *Soviet Soil Science*, **2**, 147–56.

Ruellan, A. (1968) Les horizons d'individualization et d'accumulation du calcaire dans les sols du Meroc. *Transactions of the 9th International Congress of Soil Science*, **4**, 501–10.

Ruellan, A. (1971) The history of soils: some problems of definition and interpretation, in *Paleopedology*, (ed D.H. Yaalon), International Society of Soil Science and Israel Universities Press, Jerusalem.

Ruhe, R.V. (1950) Graphic analysis of drift topographies. *American Journal of Science*, **248**, 435–43.

Ruhe, R.V. (1956) Geomorphic surfaces and the nature of soils. *Soil Science*, **82**, 441–55.

Ruhe, R.V. (1959) Stone lines in soils. *Soil Science*, **87**, 223–31.

Ruhe, R.V. (1960) Elements of the soil landscape, *Transactions of the 7th International Congress of Soil Science*, Madison, **4**, 165–9.

Ruhe, R.V. (1962) Age of the Rio Grande valley in southern New Mexico. *Journal of Geology*, **70**, 151–67.

Ruhe, R.V. (1964) Landscape morphology and alluvial deposits in southern New Mexico. *Annals of the Association of American Geographers*, **54**, 147–59.

Ruhe, R.V. (1967) Geomorphic surfaces and surficial deposits in southern New Mexico. *New Mexico Bureau of Mines and Mineral Resources Memoir*, **18**, 65pp.

Ruhe, R.V. (1969) *Quaternary landscapes in Iowa*, Iowa State University Press, Ames, Iowa.

Ruhe, R.V. (1975) *Geomorphology*, Houghton Mifflin, Boston.

Ruhe, R.V. and Olson, C.G. (1980) Soil welding. *Soil Science*, **130**, 132–9.

Runge, E.C.A. (1973) Soil development sequences and energy models. *Soil Science*, **115**, 183–93.

Russell, R.J. (1967) *River plains and sea coast*, University College of Los Angeles, Berkeley.

Rutter, N.W., Foscolos, A.E. and Hughes, O.L. (1978) Climatic trends during the Quaternary in Central Yukon based upon pedological and geomorphological evidence, in *Quaternary soils* (ed W.C. Mahaney), Geo Books, Norwich.

Ruxton, B.P. and Berry, L. (1961) Weathering profiles and geomorphic position on granite in two tropical regions. *Revue de Géomorphologie Dynamique*, **12**, 16–31.

Ruxton, B.P. and Berry, L. (1978) Clay plains and geomorphic history of the Central Sudan: A review. *Catena*, **5**, 251–84.

Ryder, J.M. (1981) Geomorphology of the southern part of the Coast Mountains, British Columbia. *Zeitschrift für Geomorphologie, Supplementband*, **37**, 120–47.

Salisbury, W. (1925) Notes on the edaphic succession in some dune soils with respect to time. *Journal of Ecology*, **13**, 322–8.

Scharpenseel, H.W. (1971) Radiocarbon dating of soils – problems, troubles, hopes, in *Paleopedology: Origin, Nature and Dating of Paleosols*, Israel University Press, Jerusalem, pp. 77–88.

Scharpenseel, H.W. and Schiffman, H. (1977) Radiocarbon dating of soils, a review. *Zeitschrift für Pflanzenernahr Bodenk.*, **140**, 159–74.

Scheidegger, A.E. (1986) The catena principle in geomorphology. *Zeitschrift für Geomorphologie*, **30**, 257–73.

Schmudde, T.H. (1963) Some aspects of landforms of the Lower Missouri River floodplain. *Annals of the Association of American Geographers*, **53**, 60–73.

Schumm, S.A. (1968) Speculations concerning paleohydrologic controls on terrestrial sedimentation. *Bulletin of the Geological Society of America*, **79**, 1573–88.

Schumm, S.A. (1972) Fluvial paleochannels, in *Recognition of ancient sedimentary environments*, Society of Economic Paleontologists and Mineralogists, Special Publication, 16, 98–107.

Schumm, S.A. (1973) Geomorphic thresholds and complex response of drainage basins, in *Fluvial geomorphology* (ed M. Morisawa), Binghamton State University of New York Publications, pp. 299–310.

Schumm, S.A. (1977) *The fluvial system*, Wiley, New York.

Schumm, S.A., Mosley, M.P. and Weaver, W.E. (1987) *Experimental fluvial geomorphology*, Wiley, New York.

Schumm, S.A. and Parker, R.S. (1973) Implications of complex response of drainage systems for Quaternary alluvial stratigraphy. *Nature*, **243**, 99–100.

Schuylenborgh, J. van (1973) Report on Topic 11: sesquioxide formation and transformation, in *Pseudogley and gley* (eds E. Schlichtung and U. Schwertmann), Verlag Chemie, Weinheim, pp. 93–102.

Scott, J.S. (1976) Geology of Canadian tills, in *Glacial till: an interdisciplinary study* (ed R.F. Legget), Royal Society of Canada, Special Publication, no. 12.

Scott, R.M. (1962) Exchangeable bases of mature well-drained soils in relation to rainfall in Eâst Africa. *Journal of Soil Science*, **13**, 1–9.

Seager, W.E. (1975) Cenozoic tectonic evolution of the Las Cruces area, New Mexico. *New Mexico Geological Society Guidebook, 26 Field Conference*, 241–50.

Seager, W.R. and Hawley, J.W. (1973) Geology of Rincon quadrangle, New Mexico. *New Mexico Bureau of Mines and Mineral Resources Bulletin*, **102**, 56pp.

Seager, W.R., Hawley, J.W. and Clemons, R.E. (1971) Geology of San Diego Mountain area, Dona Ana County, New Mexico. *New Mexico Bureau of Mines and Mineral Resources Bulletin*, **97**, 38pp.

Sehgal, J.L. and De Coninck, Fr. (1971) Identification of 14°A and 7°A clay minerals in Punjab soils. *Journal of the Indian Society of Soil Science*, **19**, 151–61.

Seppala, M. (1972) The term palsa. *Zeitschrift für Geomorphologie*, NF **16**, 463.

Seppala, M. (1983) Seasonal thawing of palsas in Finnish Lapland. *Proceedings of the 4th International Conference on Permafrost*, National Academic Press, Washington, pp. 1127–32.

Seppala, M. (1987) Periglacial phenomena of northern Fennoscandia, in *Periglacial processes and landforms in Britain and Ireland* (ed J. Boardman), Cambridge University Press, Cambridge, pp. 45–55.

Sevon, W.D. (1984) A sandstone weathering rate from northeastern Pennsylvania, *Geological Society of America Abstracts Program*, **16**, 63.

Sevon, W.D. (1989) *The rivers and valleys of Pennsylvania, then and now*, Pennsylvania Geological Survey, Harrisburg.

Sevon, W.D., Potter, N. Jr. and Crowl, G.H. (1983) Appalachian peneplains: a historical review. *Earth Science History*, **2**, 156–64.

Sharp, R.P. (1942) Soil structure in the St. Elias Range, Yukon Territory. *Journal of Geomorphology*, **5**, 274–301.

Shaw, C.F. (1930) Potent factors in soil formation. *Ecology*, **11**, 239–45.

Shields, L.M., Mitchell, C. and Drouet, F. (1957) Alga- and lichen-stabilized surface crusts as soil nitrogen sources. *American Journal of Botany*, **44**, 489–98.

Shotton, F.W. (1967) The problems and contributions of methods of absolute dating within the Pleistocene period. *Quarterly Journal of the Geological Society of London*, **122**, 357–83.

Shreve, R.L. (1966) Statistical law of stream numbers. *Journal of Geology*, **74**, 17–37.

Shroba, R.R. (1977) Soil development in Quaternary tills, rock-glacier deposits, and taluses, southern and central Rocky Mountains. Unpublished PhD thesis, University of Colorado, Boulder.

Sibirtzev, N.M. (1895) Genetic classification of soils. *Novo-Aleksandr, Agricultural Institute, Zap.*, **9**, 1–23.

Sibirtzev, N.M. (1901) Russian soil investigations. *United States Department of Agriculture Experimental Station Record*, **12**, 704–12 and 807–18.

Simonson, R.W. (1954) Identification and interpretation of buried soils. *American Journal of Science*, **252**, 705–32.

Simonson, R.W. (1959) Outline of a generalised theory of soil genesis. *Proceedings of the Soil Science Society of America*, **23**, 152–6.

Simonson, R.W. (1968) Concept of soil. *Advances in Agronomy*, **20**, 1–47.

Simonson, R.W. (1978) A multiple-process model of soil genesis, in *Quaternary Soils*, (ed W.C. Mahaney), Geo Books, Norwich, pp. 1–25.

Sinai, G., Zaslavsky, D. and Golany, P. (1981) The effect of soil surface curvature on moisture and yield – Beer Sheba observations. *Soil Science*, **132**, 367–75.

Sleeman, J.R. (1963) Cracks, peds and surfaces in some soils of the Riverine Plain, New South Wales. *Australian Journal of Soil Research*, **1**, 91–102.

Sleeman, J.R. (1964) Structure variations within two red-brown earth profiles. *Australian Journal of Soil Research*, **2**, 146–61.

Smeck, N.E. and Runge, E.C.A. (1971a) Phosphorus availability and redistribution in relation to profile development in an Illinois landscape segment. *Proceedings of the Soil Science Society of America*, **35**, 952–9.

Smeck, N.E. and Runge, E.C.A. (1971b) Factors influencing profile development exhibited by some hydromorphic soils in Illinois, in *Pseudogleys and gleys*,

258   References

International Society of Soil Science, Commission V and VI Verlag Chemie, Weinheim.

Smith, B.R. and Buol, S.W. (1968) Genesis and relative weathering studies in three semi-arid soils. *Proceedings of the Soil Science Society of America*, **32**, 261–5.

Smith, D.G. (1983) Anastomosed fluvial deposits: modern examples from Western Canada. *Special Publication of the International Association of Sedimentologists*, **6**, 155–68.

Smith, J. (1956) Some moving soils in Spitsbergen. *Journal of Soil Science*, **7**, 10–21.

Smith, R.M., Twiss, P.C., Krauss, R.K. *et al.* (1970) Dust deposition in relation to site, season and climate. *Proceedings of the Soil Science Society of America*, **34**, 112–17.

Snyder, K.E. (1988) Pedogenesis and landscape development in the Salamanca re-entrant, southwestern New York. Unpublished PhD thesis, Cornell University, Ithaca, New York.

Soil Survey Staff (1975) *Soil taxonomy – a basic system of soil classification for making and interpreting soil surveys*, United States Department of Agriculture, Soil Conservation Service, Agriculture Handbook, 436.

Sokolov, I.A. and Sokolova, T.A. (1962) Zonal soil groups in permafrost regions. *Soviet Soil Science*, **10**, 1130–6.

Soloviev, P.A. (1973) Thermokarst phenomena and landforms due to frost heaving in Central Yakutia. *Bulletyn Peryglacjalny*, **23**, 135–55.

Sparks, B.W., Williams, R.B.G. and Bell, F.G. (1972) Presumed ground-ice depressions in East Anglia. *Proceedings of the Royal Society of London*, **A327**, 329–43.

Sparrow, G.W.A. (1966) Some environmental factors in the formation of slopes. *Geographical Journal*, **132**, 390–95.

Speight, J.G. (1980) The role of topography in controlling throughflow generation: a discussion. *Earth Surface Processes*, **5**, 187–91.

Sperling, T. (1973) *The Amazon*, Time-Life International, Amsterdam.

Spinnanger, G. (1968) Global radiation and duration of sunshine in northern Norway and Spitsbergen. *Meteorologiske Annaler*, **5**. 137pp.

Stablein, G. (1977) Arktische Boden West-Gronlands, Pedovarianz in Abhangigkeit vom geookologischen Milieu. *Polarforschung*, **47**, 11–25.

Stablein, G. (1979) Boden und Relief im Westgronland. *Zeitschrift für Geomorphologie, Supplementband*, **33**, 232–45.

Stablein, G. (1984) Geomorphic altitudinal zonation in the Arctic-alpine mountains of Greenland. *Mountain Research and Development*, **4**, 319–31.

Stearns, L.A. and MacCreary, D. (1957) The case of the vanishing brick dust. *Mosquito News*, **17**, 303–4.

Steffensen, E. (1969) The climate and its recent variations at the Norwegian Arctic stations. *Meteorologiske Annaler*, **5**, 349pp.

Stephens, C.G. (1947) Functional systems in pedogenesis. *Transactions of the Royal Society of South Australia*, **71**, 168–81.

Stevens, P.R. and Walker, T.W. (1970) The chronosequence concept and soil formation. *Quarterly Review in Biology*, **45**, 333–50.

Stewart, J.E. and Bodhaine, G.L. (1961) Floods in the Skagit River Basin, Washington. *United States Geological Survey Water Supply Paper*, 1527, 66pp.

Stibbe, E. (1974) Hydrological balance of Limans in the Negev. *Volcani Institute for Agricultural Research, project* 304, 14pp.

Stork, A. (1963) Plant immigration in front of retreating glaciers with examples from the Kebnekajse area, northern Sweden. *Geografiska Annaler*, **45A**, 1–22.

Strahler, A.N. (1952) Hypsometric (area-altitude) analysis of erosional topography. *Bulletin of the Geological Society of America*, **63**, 1117–42.

Strahler, A.N. (1954) Statistical analysis in geomorphic research. *Journal of Geology*, **62**, 1–25.

Strahler, A.N. (1964) Quantitative geomorphology of drainage basins and channel networks, in *Handbook of applied hydrology* (ed V.T. Chow), McGraw-Hill, New York.

Strahler, A.N. and Strahler, A.H. (1973) *Environmental geoscience*, Hamilton, Santa Barbara, California.

Stuart, D.M., Fosberg, M.A. and Lewis, C.G. (1961) Caliche in south-western Idaho. *Proceedings of the Soil Science Society of America*, **25**, 132–5.

Sugden, D.E. and John, B.J. (1976) *Glaciers and landscape: a geomorphological approach*, Edward Arnold, London.

Summerfield, M.A. (1978) The nature and origin of silcrete with particular reference to southern Africa. Unpublished D Phil thesis, Oxford University.

Summerfield, M.A. (1983) Silcrete, in *Chemical sediments and geomorphology* (eds A.S. Goudie and K. Pye), Academic Press, London, pp. 59–91.

Svatkov, N.M. (1958) Soils of Wrangel Island. *Soviet Soil Science*, **1**, 80–87.

Swanson, D.K. (1985) Soil catenas on Pinedale and Bull Lake Moraines, Willow Lake, Wind River Mountains, Wyoming. *Catena*, **12**, 329–42.

Syers, J.K., Jackson, M.K., Berkheiser, V.E. *et al.* (1969) Eolian sediment influence on pedogenesis during the Quaternary. *Soil Science*, **107**, 421–7.

Tandarich, J.P., Darmody, R.G. and Follmer, L.R. (1987) Some connections in the history of geology and pedology as exemplified by the evolution of profile concepts, *Geological Society of America, Abstracts Program*, **19**, 864.

Tedrow, J.C.F. (1966) Polar desert soils. *Proceedings of the Soil Science Society of America*, **30**, 381–7.

Tedrow, J.C.F. (1968) Pedogenic gradients of the Polar Regions. *Journal of Soil Science*, **19**, 197–204.

Tedrow, J.C.F. (1974) Soils of the high Arctic landscapes, in *Polar deserts and modern man* (eds T.L. Smiley and J.H. Zumberge), University of Arizona Press, Tucson.

Temple, P.H. and Rapp, A. (1972) Landslides in the Mgeta area, Western Uluguru Mountains, Tanzania. *Geografiska Annaler*, **54A**, 157–93.

Ternan, J.L. and Williams, A.G. (1979) Hydrological pathways and granite weathering on Dartmoor, in *Approaches to fluvial processes* (ed A.F. Pitty), Geo Books, Norwich, pp. 5–30.

Thompson, C.H. (1983) Development and weathering of large parabolic dune systems along the subtropical coast of eastern Australia. *Zeitschrift für Geomorphologie Supplementband*, **45**, 205–25.

Thompson, C.H. and Bowman, G.M. (1984) Subaerial denudation and weathering of vegetated coastal dunes in eastern Australia, in *Coastal geomorphology in Australia* (ed B.G. Thom), Academic Press, Australia, pp. 263–90.

Thompson, M.L., Smeck, N.E. and Bigham, J.M. (1981) Parent materials and paleosols in the Teays River Valley, Ohio. *Journal of the Soil Science Society of America*, **45**, 918–25.

Thorarinsson, S. (1944) *Tefrokronologiska studier pa Island*, Ejnar Munksgaard, Copenhagen.

Thorarinsson, S. (1954) The tephra-fall from Hekla on March 29th, 1947. Pt 2, in *The eruption of Hekla, 1947–1948* (eds T. Einarsson, G. Kjartansson and S. Thorarinsson), Ejnar Munksgaard, Reykjavik.

Thorarinsson, S. (1962) L'erosion eolienne en Islande. *Revue de Geomorphologie Dynamique*, **13**, 107–24.

Thorarinsson, S. (1970) Tephrochronology and medieval Iceland, in *Scientific techniques in medieval archaeology* (ed R. Beyer), University of California Press, Los Angeles, pp. 295–328.

Thorn, C.E. (1976) Quantitative evolution of nivation in the Colorado Front Range. *Bulletin of the Geological Society of America*, **87**, 1169–78.

Thorn, C.E. (1979) Ground temperatures and surficial transport in colluvium during snowpatch meltout, Colorado Front Ranges. *Arctic and Alpine Research*, **11**, 41–52.

Thorn, C.E. (1988) Nivation: a geomorphic chimera, in *Advances in periglacial geomorphology* (ed M.J. Clark), Wiley, Chichester, pp. 3–31.

Thornbury, W.B. (1954) *Principles of geomorphology*, Wiley, New York.

Thornbury, W.B. (1965) *Regional geomorphology of the United States*, Wiley, New York.

Tonkin, P.J. (1985) Studies of soil development with respect to aspect and rainfall, eastern hill country, South Island, New Zealand, in *Proceedings of the Soil Dynamics and Land Use Seminar, Blenheim, May 1985* (ed I.B. Campbell), New Zealand Society of Soil Science and New Zealand Soil Conservation Association, pp. 1–18.

Tonkin, P.J. and Basher, L.R. (1990) Soil-stratigraphic techniques in the study of soil and landform evolution across the Southern Alps, New Zealand. *Geomorphology*, **3**, 547–75.

Tonkin, P.J., Harrison, J.B.J., Whitehouse, I.E. *et al.* (1981) Methods for assessing late Pleistocene and Holocene erosion history in glaciated mountain drainage basins, in *Erosion and sediment transport in Pacific Rim steeplands* (eds T.R.H. Davies and A.J. Pierce), International Association of Scientific Hydrology, 132, 527–40.

Trendall, A.F. (1962) The formation of apparent peneplains by a process of combined lateritisation and surface wash. *Zeitschrift für Geomorphologie*, **NS 6**, 183–97.

Tricart, J. (1970) *Geomorphology of cold environments*, (trans.), McMillan, New York.

Tricart, J. (1972) *Landforms of the humid tropics, forests and savannas*, Longman, London.

Tricart, J. and Cailleux, A. (1967) Le modèle des régions périglaciaires, *Traite de géomorphologie*, **11**, SEDES, Paris, 512pp.

Tricart, J. and Cailleux, A. (1972) *Climatic geomorphology*, Longman, London.

Troeh, F.R. (1964) Landform parameters correlated to soil drainage. *Proceedings of the Soil Science Society of America*, **28**, 808–12.

Troll, C. (1944) Struckturboden, solifluktion und frostklimate der erde. *Geologische Rundschau*, **34**, 545–694.

Tsyganenko, A.F. (1968) Aeolian migration of water soluble matter and the probable geochemical and soil formation significance. *Transactions of the 9th International Congress of Soil Science*, **4**, 333–42.

Twidale, C.R. (1962) Steepened margins of inselbergs from north-western Eyre Peninsula, South Australia. *Zeitschrift für Geomorphologie*, **NF 6**, 51–69.

Ugolini, F.C. (1966) Soils of the Mesters Vig District, northeast Greenland. II. Exclusive of Arctic Brown and Podzol-like soils, *Meddeleser om Gronland*, **176**, 25pp.

Ugolini, F.C. (1968) Soil development and alder invasion in a recently deglaciated area of Glacier bay, Alaska, in *The biology of alder* (eds J.M. Trappe *et al.*), United States Forestry Service, Portland, pp. 115–40.

Valentine, K.W.G. and Dalrymple, J.B. (1976) Quaternary buried paleosols: a critical review. *Quaternary Research*, **6**, 209–22.

Valentine, K.W.G., Fladmark, K.P. and Spurling, B.E. (1980) The description, chronology and correlation of buried soils and cultural layers in a terrace section, Peace River Valley, British Columbia. *Canadian Journal of Soil Science*, **60**, 185–98.

Van der Sluijs, P. (1970) Decalcification of marine clay soils connected with decalcification during silting. *Geoderma*, **4**, 209–27.

Vanoni, V.A. (1971) Sediment transportation mechanics question: genetic classification of valley sediment deposits. *Journal of the Hydraulics Division of the American Society of Civil Engineers*, **95**, HYI, 43–53.

Van Straaten, L.M.J.U. (1954) Composition and structure of recent marine sediments in the Netherlands. *Leidse Geol. Mededel.*, **10**, 58–71.

Van Vliet-Lanoe, B. (1976) Traces de segregation de glace en lentilles associees aux sols et phenomenes periglaciaires fossiles. *Bulletyn Periglacjalny*, **26**, 41–55.

Van Vliet-Lanoe, B. (1985) Frost effects in soils, in *Soils and Quaternary Landscape Evolution* (ed J. Boardman), Wiley, Chichester, pp. 117–58.

Van Vliet-Lanoe, B and Langohr, R. (1981) Correlation between fragipans and permafrost with special reference to Weichsel silty deposits in Belgium and northern France. *Catena*, **8**, 137–54.

Veneman, P.L.M. and Bodine, S.M. (1982) Chemical and morphological soil characteristics in a New England drainage-toposequence. *Journal of the Soil Science Society of America*, **46**, 359–63.

Veneman, P.L.M., Vepraskas, M.J. and Bouma, J. (1976) The physical significance of soil mottling in a Wisconsin toposequence. *Geoderma*, **15**, 103–18.

Vepraskas, M.J. and Bouma, J. (1976) Model experiments on mottle formation simulating field conditions. *Geoderma*, **15**, 217–30.

Verhoeven, B. (1962) On the calcium carbonate content of young marine sediments. *Netherlands Journal of Agricultural Science*, **10**, 58–71.

Vine, H. (1941) A soil catena in the Nigerian Cocoa Belt. *Farm and Forest*, **2**, 139–41.

Vine, H. *et al.* (1954) Progress of soils surveys in south-west Nigeria. *Proceedings of the 2nd Inter-African Soil Conference*, 211–36.

Vreeken, W.J. (1975) Principal kinds of chronosequences and their significance in soil history. *Journal of Soil Science*, **26**, 378–94.

Vreeken, W.J. (1984) Soil-landscape chronograms for pedochronological analysis. *Geoderma*, **34**, 149–64.

Walker, P.H. (1962a) Soil layers on hillslopes: a study at Nowra, New South Wales, Australia. *Journal of Soil Science*, **13**, 167–77.

Walker, P.H. (1962b) Terrace chronology and soil formation on the south coast of New South Wales. *Journal of Soil Science*, **13**, 178–86.

Walker, P.H. (1966) *Postglacial environments in relation to landscape and soils on the Cary Drift, Iowa*, Iowa State University Agriculture and Home Economics Experimental Station Research Bulletin, 549, 835–75.

Walker, P.H. and Coventry, R.J. (1976) Soil profile development in some alluvial deposits of Eastern New South Wales. *Australian Journal of Soil Research*, **14**, 305–17.

Walker, P.H. and Ruhe, R.V. (1968) Hillslope models and soil formation. II. Closed systems. *Transactions of the 9th International Conference on Soil Science*, **4**, 561–8.

Walker, P.H., Hall, G.F. and Protz, R. (1968a) Soil trends and variability across selected landscapes in Iowa. *Proceedings of the Soil Science Society of America*, **32**, 97–101.

Walker, P.H., Hall, G.F. and Protz, R. (1968b) Relation between landform parameters and soil properties. *Proceedings of the Soil Science Society of America*, **32**, 101–4.

Washburn, A.L. (1956) Classification of patterned ground and review of suggested origins. *Bulletin of the Geological Society of America*, **67**, 823–56.

Washburn, A.L. (1973) *Periglacial processes and environments*, Edward Arnold, London.

Washburn, A.L. (1979) *Geocryology: a survey of periglacial processes and environments*, Wiley, London.

Watson, A. (1982) The origin, nature and distribution of gypsum crusts in deserts. Unpublished D. Phil. thesis, University of Oxford.

Watson, E. (1971) Remains of pingos in Wales and the Isle of Man. *Geological Journal*, **7**, 381–92.

Watson, J.P. (1964–5) A soil catena on granite in southern Rhodesia. *Journal of Soil Science*, **15**, 238–57; **16**, 158–69.

Watson, J.P. (1965) Soil catenas. *Soils and Fertiliser*, **28**, 307–10.

Watts, N.L. (1980) Quaternary pedogenic calcretes from the Kalahari (southern Africa): mineralogy, genesis and diagenesis. *Sedimentology*, **27**, 661–86.

Watts, W.A. (1979) Late Quaternary vegetation of central Appalachia and the New Jersey coastal plain. *Ecological Monographs*, **49**, 427–69.

Wayland, E.J. (1921) *A general account of the geology of Uganda by the geologist.* Report of the Geological Department of Uganda (1921), 8–20.

Wayland, E.J. (1933) The peneplains of East Africa. *Geographical Journal*, **82**, 95.

Wayland, E.J. (1934) The peneplains of East Africa. *Geographical Journal*, **83**, 79.

Webster, R. (1965) A catena of soils on the Northern Rhodesia plateau. *Journal of Soil Science*, **16**, 31–43.

Webster, R. (1973) Automatic soil boundary location from transect data. *Mathematical geology*, **5**, 27–37.

Webster, R. and De La Cuanalo, C.H.E. (1975) Soil transect correlograms of North Oxfordshire and their interpretation. *Journal of Soil Science*, **26**, 176–94.

Webster, R. and Oliver, M.A. (1990) *Statistical methods of soil and resource survey*, Oxford University Press, Oxford.

Weidner, E. (1981) Geomorphologisch bedingte Differenzierungen der Bodengesellschaften am Sudabfall des Himalayas (Sud-Nepal). *Zeitschrift für Geomorphologie, Supplementband*, **39**, 123–37.

Weihaupt, J.G. (1977) Morphometric definitions and classifications of oxbow lakes, Yukon River Basin, Alaska. *Water Resources Research*, **13**, 195–6.

Weldon, R.J. (1986) Late Cenozoic geology of Cajon Pass: implications for tectonics and sedimentation along the San Andreas fault, unpublished PhD thesis, California Institute of Technology.

Wells, S.G. and McFadden, L.D. (1987) Influence of Late Quaternary climatic changes on geomorphic and pedogenic processes on a Desert Piedmont, Eastern Mojave Desert, California. *Quaternary Research*, **27**, 130–46.

Wells, S.G., Dohrenwend, J.C., McFadden, L.D. *et al.* (1985) Late Cenozoic landscape evolution of lava flow surfaces of the Cima volcanic field, Mojave Desert, California. *Bulletin of the Geological Society of America*, **96**, 1518–29.

Whipkey, R.Z. (1965) Subsurface stormflow from forested slopes. *Bulletin of the International Association of Scientific Hydrology*, **10**, 74–85.

Whitfield, W.A.D. and Furley, P.A. (1971) The relationship between soil patterns and slope form in the Ettrick Association, south-east Scotland, in *Slopes, form and process* (ed D. Brunsden), Institute of British Geographers Special Publication, no. 3, 165–75.

Whittaker, R.H., Buol, S.W., Niering, W.A. *et al.* (1968) A soil and vegetation pattern in the Santa Catalina Mountains, Arizona. *Soil Science*, **105**, 440–51.

Wilde, S.A. (1946) *Forest soils and forest growth*, Walthur, Chronica Botanica.

Wilding, L.P. (1967) Radiocarbon dating of biogenic opal. *Science*, **156**, 66–7.

Williams, A.G., Ternan, J.L. and Kent, M. (1984) Hydrochemical characteristics of a Dartmoor hillslope, in *Catchment experiments in fluvial geomorphology* (eds T.P. Burt and D.E. Walling), Geo Books, Norwich, pp. 379–98.

Williams, M.A.J. (1968) A dune catena on the clay plains of the West Central Gezira, Republic of the Sudan. *Journal of Soil Science*, **19**, 367–78.

Wilson, R.C.L. (ed) (1983) *Residual deposits: surface related weathering processes and materials*, Blackwell Scientific Publishers, Oxford.

Wintle, A.G. (1981) Thermoluminescence dating of late Devensian loesses in southern England. *Nature*, **289**, 479–80.

Wintle, A.G. and Catt, J.A. (1985) Thermoluminescence dating of soils developed in Late Devensian loess at Pegwell Bay, Kent. *Journal of Soil Science*, **36**, 293–8.

Wintle, A.G. and Huntley, D.J. (1982) Thermoluminescence dating of sediments. *Quaternary Science Reviews*, **1**, 31–53.

Wintle, A.G., Shackleton, N.J. and Lautridou, J.P. (1984) Thermoluminescence dating of periods of loess deposition and soil formation in Normandy. *Nature*, **310**, 491–3.

Wischmeier, W.H. (1975) Cropland erosion and sedimentation, in *Control of water pollution from cropland, Vol. II. An overview*, Agricultural Research Service and Environmental Protection Agency, Washington.

Woodcock, A.H. (1974) Permafrost and climatology of a Hawaii volcano crater. *Arctic and Alpine Research*, **6**, 49–62.

Woolnough, W.G. (1918) The physiographic significance of laterite in Western Australia. *Geological Magazine*, **5**, 385–93.

Woolnough, W.G. (1927) The duricrust of Australia. *Journal of the Royal Society of New South Wales*, **61**, 24–53.

Working Group on the Origin and Nature of Paleosols (1971), Report, in *Paleopedology*, (ed D.H. Yaalon), International Society of Soil Science and Israel Universities Press, Jerusalem.

Yaalon, D.H (1963) On the origin and accumulation of salts in groundwater and soils in Israel. *Bulletin of the Research Council of Israel*, **11G**, 105–31.

Yaalon, D.H. (1971) Soil-forming processes in time and space, in *Paleopedology* (ed D.H. Yaalon), International Society of Soil Science and Israel Universities Press, Jerusalem.

Yaalon, D.H. (1975) Conceptual models in pedogenesis: can soil forming factors be solved? *Geoderma*, **14**, 189–205.

Yaalon, D.H. and Lomas, J. (1970) Factors controlling the supply and the chemical composition of aerosols in a near-shore and coastal environment. *Agricultural Meteorology*, **7**, 443–54.

Yair, A. (1983) Hillslope hydrology, water harvesting and areal distribution of some ancient agricultural systems, northern Negev, Israel. *Journal of Arid Environments*, **6**, 283–301.

Yair, A. (1990) The role of topography and surface cover upon soil formation along hillslopes in arid climates. *Geomorphology*, **3**, 287–99.

Yair, A. and Danin, A. (1980) Spatial variations in vegetation as related to the soil moisture regime over an arid limestone hillside, northern Negev, Israel. *Oecologia, (Berlin)*, **47**, 83–88.

Yair, A. and De Ploey, J. (1979) Field observations and laboratory experiments concerning the creep process of rock blocks in an arid environment. *Catena*, **6**, 245–58.

Yair, A. and Lavee, H. (1985) Runoff generation in arid and semi-arid zones, in *Hydrological forecasting* (eds M.G. Anderson and T.P. Burt), Wiley, Chichester, pp. 183–220.

York, J.C. and Dick-Peddie, W.A. (1969) Vegetation changes in southern New Mexico during the past hundred years, in *Arid lands in perspective* (eds W.G. McGinnies and B.J. Goldman), University of Arizona Press, pp. 155–66.

Young, A. (1958) Some considerations of slope form and development, regolith and denudational processes, unpublished PhD thesis, University of Sheffield.

Young, A. (1960) Soil movement by denudation processes on slopes. *Nature*, **188**, 120–22.

Young, A. (1963) Deductive models of slope evolution. *Neue Beitrage zur internationalen Hangforschung*, Vandenhoeck and Ruprecht, Gottingen.

Young, A. (1968) Slope form and the soil catena in savanna and rainforest environments. *British Geomorphological Research Group, Occasional Publication*, **5**, 3–12.

Young, A. (1969) The accumulation zone on slopes. *Zeitschrift für Geomorphologie*, **NF 13**, 231–3.

Young, A. (1971) Slope profile analysis: the system of best units. *Institute of British Geographers Special Publication*, **3**, 1–13.

Young, A. (1972a) *Slopes*, Oliver and Boyd, Edinburgh.

Young, A. (1972b) The soil catena: a systematic approach, in *International Geography 1972*, (eds W.P. Adams and F.M. Helleiner), International Geography, Toronto.

Young, A. (1976) *Tropical Soils and Soil Survey*, Cambridge University Press, Cambridge.

Young, A. and Stephen, I. (1965) Rock weathering and soil formation on high altitude plateaux of Malawi. *Journal of Soil Science*, **16**, 323–33.

Zaslavsky, D. and Rogowski, A. (1969) Hydrologic and morphologic implications of anisotropy and inflitration in soil profile development. *Proceedings of the Soil Science Society of America*, **33**, 594–9.

Zonneveld, I.S. (1960) De Branantse Biesboch: a study of soil and vegetation of a fresh water tidal delta. *Stichting voor Bodemkartering Wageningen – Bodemkundige Studies*, vol. 4.

Zuur, A.J. (1936) *Over de bodemkundige gesteldheid van de Wieringermeer*, Algemeene Landsdrukkeriz, The Hague,

Index

Page numbers in **bold** refer to figures and those in *italic* refer to tables.

Soil Geomorphology
An integration of pedology and geomorphology
John Gerrard

Soil formation is related to the nature of landforms on which that formation takes place. The nature of the soil affects some of the processes shaping the landforms and which cause landscape change. An accurate assessment of the evolution of landforms, and of the patterns of soil formation, is possible only if the interdependence of soils and landforms is recognized. This book provides an integration of geomorphology and pedology to fully assess this relationship.

After an opening chapter outlining the approach adopted, specific chapters examine the relationship between soil formation and specific suites of landform assemblages. The highly influential catena concept is followed by an analysis of soils within drainage holes, on erosion surfaces, flood plains, river terraces, coastal plains, glacial and fluvioglacial landforms, deserts, and periglacial landforms. The concluding chapter addresses the way in which soils can be used to assess major phases in landscape evolution.

Undergraduate students and lecturers in geography or soil sciences will find this a valuable interdisciplinary introduction to soil geomorphology.

John Gerrard is Senior Lecturer in Geography, School of Geography, University of Birmingham, UK.

Also available

Geomorphological Techniques
Edited by A. Goudie, M. Anderson, T. Burt, J. Lewin, K. Richards, B. Whalley and P. Worsley
2nd edn, paperback (0 0 4445715 4), 568 pages

An Introduction to Theoretical Geomorphology
C.E. Thorn
Hardback (0 0 4551117 9) and paperback (0 0 4551118 7), 264 pages

History of Geomorphology
From Hutton to Hack
Edited by K.J. Tinkler
Hardback (0 0 4551138 1), 368 pages

CHAPMAN & HALL
London · Glasgow · New York · Tokyo · Melbourne · Madras